T0258063

Advanced Processing and Manufacturing Technologies for Structural and Multifunctional Materials VI

Advanced Processing and Manufacturing Technologies for Structural and Multifunctional Materials VI

A Collection of Papers Presented at the 36th International Conference on Advanced Ceramics and Composites January 22–27, 2012 Daytona Beach, Florida

Edited by
Tatsuki Ohji
Mrityunjay Singh

Volume Editors
Michael Halbig
Sanjay Mathur

A John Wiley & Sons, Inc., Publication

Published by John Wiley & Sons, Inc., Hoboken, New Jersey.
Published simultaneously in Canada.

For general information on our other products and services or for technical support, please contact our
Customer Care Department within the United States at (800) 762-2974, outside the United States at
(317) 572-3993 or fax (317) 572-4002.

Wiley also publishes its books in a variety of electronic formats. Some content that appears in print may
not be available in electronic formats. For more information about Wiley products, visit our web site at
www.wiley.com.

Library of Congress Cataloging-in-Publication Data is available.

ISBN: 978-1-118-20598-3
ISSN: 0196-6219

Printed in the United States of America.

10 9 8 7 6 5 4 3 2 1

Contents

Preface

The Sixth International Symposium on Advanced Processing and Manufacturing Technologies for Structural and Multifunctional Materials and Systems (APMT) was held during the 36th International Conference on Advanced Ceramics and Composites, in Daytona Beach, FL, January 22 - 27, 2012. The aim of this international symposium was to discuss global advances in the research and development of advanced processing and manufacturing technologies for a wide variety of non-oxide and oxide based structural ceramics, particulate and fiber reinforced composites, and multifunctional materials. This year's symposium also honored Professor R. Judd Diefendorf, Clemson University, USA, recognizing his outstanding contributions to science and technology of advanced ceramic fibers, carbon-carbon, and ceramic matrix composites and his tireless efforts in promoting their wide scale industrial applications. A total of 68 papers, including invited talks, oral presentations, and posters, were presented from 15 countries (USA, Japan, Germany, China, Korea, France, Australia, Denmark, Estonia, India, Italy, Luxembourg, Serbia, Slovenia, and Taiwan). The speakers represented universities, industry, and research laboratories.

This issue contains 17 invited and contributed papers, all peer reviewed according to The American Ceramic Society review process. The latest developments in processing and manufacturing technologies are covered, including green manufacturing, smart processing, advanced composite manufacturing, rapid processing, joining, machining, and net shape forming technologies. These papers discuss the most important aspects necessary for understanding and further development of processing and manufacturing of ceramic materials and systems.

The editors wish to extend their gratitude and appreciation to all the authors for their cooperation and contributions, to all the participants and session chairs for their time and efforts, and to all the reviewers for their valuable comments and suggestions. Financial support from the Engineering Ceramic Division and The American Ceramic Society is gratefully acknowledged. Thanks are due to the staff of the meetings and publication departments of The American Ceramic Society for their invaluable assistance.

We hope that this issue will serve as a useful reference for the researchers and technologists working in the field of interested in processing and manufacturing of ceramic materials and systems.

TATSUKI OHJI, Nagoya, Japan
MRITYUNJAY SINGH, Cleveland, OH, USA

Introduction

This issue of the Ceramic Engineering and Science Proceedings (CESP) is one of nine issues that has been published based on content presented during the 36th International Conference on Advanced Ceramics and Composites (ICACC), held January 22–27, 2012 in Daytona Beach, Florida. ICACC is the most prominent international meeting in the area of advanced structural, functional, and nanoscopic ceramics, composites, and other emerging ceramic materials and technologies. This prestigious conference has been organized by The American Ceramic Society's (ACerS) Engineering Ceramics Division (ECD) since 1977.

The 36th ICACC hosted more than 1,000 attendees from 38 countries and had over 780 presentations. The topics ranged from ceramic nanomaterials to structural reliability of ceramic components which demonstrated the linkage between materials science developments at the atomic level and macro level structural applications. Papers addressed material, model, and component development and investigated the interrelations between the processing, properties, and microstructure of ceramic materials.

The conference was organized into the following symposia and focused sessions:

Symposium 1	Mechanical Behavior and Performance of Ceramics and Composites
Symposium 2	Advanced Ceramic Coatings for Structural, Environmental, and Functional Applications
Symposium 3	9th International Symposium on Solid Oxide Fuel Cells (SOFC): Materials, Science, and Technology
Symposium 4	Armor Ceramics
Symposium 5	Next Generation Bioceramics

Symposium 6	International Symposium on Ceramics for Electric Energy Generation, Storage, and Distribution
Symposium 7	6th International Symposium on Nanostructured Materials and Nanocomposites: Development and Applications
Symposium 8	6th International Symposium on Advanced Processing & Manufacturing Technologies (APMT) for Structural & Multifunctional Materials and Systems
Symposium 9	Porous Ceramics: Novel Developments and Applications
Symposium 10	Thermal Management Materials and Technologies
Symposium 11	Nanomaterials for Sensing Applications: From Fundamentals to Device Integration
Symposium 12	Materials for Extreme Environments: Ultrahigh Temperature Ceramics (UHTCs) and Nanolaminated Ternary Carbides and Nitrides (MAX Phases)
Symposium 13	Advanced Ceramics and Composites for Nuclear Applications
Symposium 14	Advanced Materials and Technologies for Rechargeable Batteries
Focused Session 1	Geopolymers, Inorganic Polymers, Hybrid Organic-Inorganic Polymer Materials
Focused Session 2	Computational Design, Modeling, Simulation and Characterization of Ceramics and Composites
Focused Session 3	Next Generation Technologies for Innovative Surface Coatings
Focused Session 4	Advanced (Ceramic) Materials and Processing for Photonics and Energy
Special Session	European Union – USA Engineering Ceramics Summit
Special Session	Global Young Investigators Forum

The proceedings papers from this conference will appear in nine issues of the 2012 Ceramic Engineering & Science Proceedings (CESP); Volume 33, Issues 2-10, 2012 as listed below.

- Mechanical Properties and Performance of Engineering Ceramics and Composites VII, CESP Volume 33, Issue 2 (includes papers from Symposium 1)
- Advanced Ceramic Coatings and Materials for Extreme Environments II, CESP Volume 33, Issue 3 (includes papers from Symposia 2 and 12 and Focused Session 3)
- Advances in Solid Oxide Fuel Cells VIII, CESP Volume 33, Issue 4 (includes papers from Symposium 3)
- Advances in Ceramic Armor VIII, CESP Volume 33, Issue 5 (includes papers from Symposium 4)

- Advances in Bioceramics and Porous Ceramics V, CESP Volume 33, Issue 6 (includes papers from Symposia 5 and 9)
- Nanostructured Materials and Nanotechnology VI, CESP Volume 33, Issue 7 (includes papers from Symposium 7)
- Advanced Processing and Manufacturing Technologies for Structural and Multifunctional Materials VI, CESP Volume 33, Issue 8 (includes papers from Symposium 8)
- Ceramic Materials for Energy Applications II, CESP Volume 33, Issue 9 (includes papers from Symposia 6, 13, and 14)
- Developments in Strategic Materials and Computational Design III, CESP Volume 33, Issue 10 (includes papers from Symposium 10 and from Focused Sessions 1, 2, and 4)

The organization of the Daytona Beach meeting and the publication of these proceedings were possible thanks to the professional staff of ACerS and the tireless dedication of many ECD members. We would especially like to express our sincere thanks to the symposia organizers, session chairs, presenters and conference attendees, for their efforts and enthusiastic participation in the vibrant and cutting-edge conference.

ACerS and the ECD invite you to attend the 37th International Conference on Advanced Ceramics and Composites (http://www.ceramics.org/daytona2013) January 27 to February 1, 2013 in Daytona Beach, Florida.

MICHAEL HALBIG AND SANJAY MATHUR
Volume Editors
July 2012

CONTRIBUTION TO THE UNDERSTANDING OF THE MICROSTRUCTURE OF FIRST GENERATION SI-C-O FIBERS

Francis Teyssandier, Géraldine Puyoo, Stéphane Mazerat, Georges Chollon, René Pailler, Florence Babonneau[1]
Laboratoire des composites thermostructuraux, Université de Bordeaux 1, Pessac, France.
1 Chimie de la Matière Condensée de Paris, Collège de France, Paris, France.

ABSTRACT:
As compared to the most recent SiC-based ceramic fibers, first generation fibers include a significant amount of oxygen and free carbon. Though a large number of papers have been devoted to understanding the microstructure of these fibers, their composition and microstructure are still controversial. This communication is intended to propose a microstructure description of these fibers according to their composition in the Si-C-O isothermal section of the phase diagram. The proposed microstructure is deduced from a large set of characterizations including XRD, Raman spectroscopy, RMN, and elemental analysis. Three Nicalon and four Tyranno fibers are thus characterized. Their fracture behavior is also described.

KEYWORDS: Si-C-O fibers, Fiber composition, Fiber microstructure.

INTRODUCTION

Ceramic matrix composites (CMC) were first developed for applications in severe environments: at high temperature, under mechanical stress, in oxidative environments and even under radiations. Such demanding applications require very high properties that are obtained by use of high performance materials. New trends are to develop CMC materials for very long life time applications but at lower temperatures and with a reduced stress level. Such CMC that are aimed at being used in civil aeronautics (aircraft engine plug or exhaust…), do not require the most capable third generation fibers. Instead, the less demanding parts in terms of performance can be designed with less expensive fibers of the first generation. The aim of the present paper is to compare the properties of a variety of first generation silicon carbide fibers.

Since the pioneering work of Yajima et al.[1] in the seventies, the SiC fibers have been improved in order to increase the creep and oxidation resistance[2]. Three families of SiC fibers have been successively developed from the first generation that included carbon and oxygen in excess, to the third generation, which is composed of almost pure silicon carbide. The durability of SiC/SiC composites is at last related to the fibers lifetime, i.e., their sensitivity to oxidation or subcritical crack growth. This latter mechanism is largely influenced by the composition and the structure of the Si-O-C fibers and is expected to be accelerated by the presence of free carbon. The percentage of oxygen and the SiC grain size may also have a significant influence on the fiber reactivity. We therefore undertook the characterization of a wide variety of first generation silicon carbide fibers in order to provide a detailed description of their microstructure and properties.

The microstructure of first generation SiC fibers is globally known. It can be described as a continuum consisting of glassy silicon oxycarbide binding pure β-SiC nanocrystals and including free carbon nanodomains called basic structural units (BSU)[3-8].

Many authors have contributed to the assessment of the microstructure of the various phases composing the Si-O-C fibers. In 1989, Laffon et al.[3] studied the NicalonTM NG100 and NG200 fibers mainly by EXAFS and X-ray diffraction and proposed a model of structure including a continuum of β-SiC nanocrystals embedded in a Si-C-O glass consisting of mixed SiO_XC_Y (X+Y=4) tetrahedral environments. Carbon BSU are described as aggregates composed of graphene layers with edges saturated with hydrogen, i.e. without any chemical bonding with the Si-O-C continuum. In 1993, Le

1

Coustumer et al.[5] studied the Nicalon fiber NLM202. They determined the following composition: 55wt% of β-SiC crystals, 40wt% of amorphous $SiO_{1.15}C_{0.85}$ and 5wt% of free carbon. Knowing the percentage of free carbon and the atomic chemical composition, they calculated the proportion of each phase. An average grain size of 1.6nm was measured using (111)-SiC Dark Field transmission imaging (DF-TEM).

In 1996 Bodet et al.[9] calculated phase proportions and composition of the fiber Nicalon NLP201 knowing atomic chemical composition and SiC/SiO_xC_y ratio as determined by XPS analysis (from the various Si2p peak components). An average β-SiC grain size of 2.7 nm was measured using (111)-SiC DF-TEM image.

More recently, by use of density functional theory, P. Kroll calculated the structure and properties of amorphous silicon oxycarbide glasses, pure[10] or including a free carbon phase[11].

RESULTS AND DISCUSSION

We studied the following first generation SiC fibers: Nicalon NLM202 and 207 from Nippon Carbon, and Tyranno S, ZMI, LoxM from UBE. We also studied the Tyranno AM and Nicalon NLP101 which are not currently commercialized. Their composition and structure are nevertheless interesting to compare to other fibers.

The atomic composition of the whole fiber was first determined by chemical analysis. We then successively studied the β-SiC nanocrystallites by both X-ray diffraction (XRD) and high resolution transmission electron microscopy (HRTEM). The pure Si-C-O glass phase has the composition $SiC_xO_{2(1-x)}$. Its composition was both deduced from the atomic composition and by nuclear magnetic resonance (NMR) ^{29}Si analysis. The amount of free carbon was both deduced from the difference between the atomic composition of the fiber and the composition of the Si-C-O continuum, as well as from ^{13}C-NMR analysis. We also characterized the in-plane disorder of carbon BSU by Raman spectroscopy.

Elemental Composition of Fibers

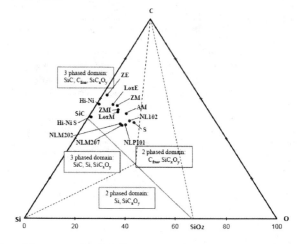

Figure 1: composition of first generation SiC fibers presented in the ternary section of the Si-O-C phase diagram.

The elemental composition of fibers was measured by chemical analysis at the CNRS facility in Solaize. The content of silicon, as well as of Zr, Ti and Al (that are present in limited amounts), was determined by Inductively Coupled Plasma - Atomic Emission Spectroscopy. The carbon concentration was deduced from the amount of CO_2 resulting from oxidation. The oxygen concentration of fibers was deduced from the amount of carbon monoxide formed by reaction at high temperature between ground fibers and the carbon crucible in which they are disposed. Hydrogen was deduced from the amount of water vapor formed by oxidation. The compositions of the fibers are plotted in the isothermal Si-C-O ternary section in figure 1. The Tyranno S fiber has the highest oxygen content and also the largest amount of amorphous $SiC_xO_{2(1-x)}$ phase.

Phase Composition of the Fibers: Assumptions

The pure Si-C-O amorphous phase which is composed of mixed tetrahedral environments SiO_XC_Y (X+Y=4) can be described using a rule of mixture between silica and silicon carbide similar to: $SiC_xO_{2(1-x)}$= x SiC + (1-x) SiO_2. We assumed that the so called "free carbon" phase, embedded into the continuum is not bonded to the amorphous $SiC_xO_{2(1-x)}$ glass nor the β-SiC grains[7]. We could thus infer that all the oxygen atoms are bonded to silicon, the remaining silicon being bonded to carbon. Carbon atoms that are not bonded to sp^3-silicon are hence part of the free carbon domains (sp^2-C).

Composition of the Si-C-O Continuum

Composition of the Si-C-O continuum (β-SiC + amorphous silicon oxycarbide) was deduced from the atomic composition measured by chemical analysis on each type of fiber. It was calculated according to the formula $SiC_xO_{2(1-x)}$ on grounds of assumptions described in the preceding paragraph.

The composition was also measured by nuclear magnetic resonance (NMR) analysis of silicon (^{29}Si) carried out on grounded fibers. NMR analysis enabled us to characterize quantitatively the proportion of the various allowed tetrahedral environments of silicon: $\underline{Si}C_4$, $\underline{Si}C_3O$, $\underline{Si}C_2O_2$, $SiCO_3$ and SiO_4. The recorded spectra were decomposed into five peaks, one for each silicon environment, the amount of each environment being proportional to the surface of the corresponding peak. The related chemical shifts are observed in the following order: δ($\underline{Si}C_3O$)> δ($\underline{Si}C_4$)> δ($\underline{Si}C_2O_2$> δ($\underline{Si}O_3C$)> δ($\underline{Si}O_4$). It is interesting to notice that the chemical shift of the $\underline{Si}C_3O$ environment is not, as expected, located between those of SiC_4 and SiC_2O_2, due to a non-linear dependence of the chemical shift (δ) to the charge (q) of atoms.

Compositions obtained by both approaches are in good accordance (figure 2). This result validates our assumptions and especially the absence of bonding between BSU and their neighboring phases.

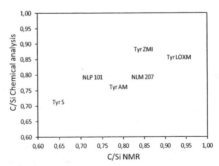

Figure 2: C/Si ratio in the Si-C-O continuum of the various fibers: comparison between calculations based on atomic composition and NMR measurements

NMR measurements revealed that both Nicalon NLP101 and NLM207, as well as Tyranno S and AM include a significant amount of SiO_4 environments (11% for Tyranno S and Nicalon NLP101). This observation provides evidence of a bulk silicon oxycarbide phase between the β-SiC grains and the BSU.

SiC Characterization by X-Ray Diffraction

The amount of crystallized β-SiC phase included in each fiber was determined by X-ray diffraction, carried out on grounded fibers, and including an internal silicon standard. The quantitative analysis was performed by comparing the areas of SiC and Si peaks after calibration with a known mixture. The error of such a procedure was estimated at 4 wt%.

The mean size of SiC crystallites was assessed by the Scherrer formula. As the basic formula assumes monodisperse spherical particles, we used a modified formula including a corrective factor[12-13] to account for the real crystals:

$$L = \frac{k\lambda}{D\cos\theta} \qquad (1)$$

L is the full-width half-maximum (FWHM) of the SiC peak, λ=0.15406 nm Cu-Kα wavelength, k a constant determined to be 0.9 by comparison with crystallite size as observed by TEM and D the mean size of the SiC crystallites.

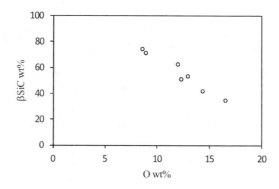

Figure 3: β-SiC crystallite size as a function of the oxygen wt%

The amount of β-SiC crystallites is clearly correlated to the oxygen content of the fiber: the higher the oxygen content, the less the amount of β-SiC crystallites (figure 3). This behavior is in accordance with the Si-C-O phase diagram. It was furthermore observed that the size of the β-SiC crystallites increases with their amount.

Determination of the Free Carbon Content

The amount of free carbon embedded into the continuum was deduced from the atomic composition of the fiber, according to the above described calculation. It was also measured by NMR ^{13}C analysis. The recorded spectrum is composed of two broad peaks at ~19ppm and 119ppm which respectively correspond to C sp3 and C sp2. From the proportion of each type of carbon hybridization, it is possible to estimate the amount of carbon bonded to silicon and the amount of free carbon.

The amount of free carbon measured by the two approaches are in good accordance (figure 4) and shows the same order among fibers. It can nevertheless be observed that amount of free carbon deduced from the measured atomic composition is always higher than the amount measured by NMR. As a general trend, carbon content of Tyranno fibers is much higher than that of Nicalon fibers, because of the type of polymer used as precursor. The maximum carbon content is measured in the Tyranno S fiber and the minimum in the Nicalon NLM207.

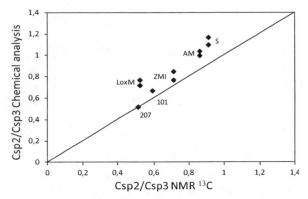

Figure 4: comparison between the ratios between free and tetrahedral carbon deduced from the measured atomic composition of the fibers and its measurement by NMR 13C.

Fractographic Examination of Rupture Patterns

The mechanical properties of fibers are strongly dependent on their chemical composition. For instance, first generation Si-C-O fibers have a tensile modulus of about 200 GPa[6], though the modulus of near stoichiometric fibers reaches almost 300-400 GPa[14]. The morphology of the rupture surface, which is typical of brittles materials, is indicative of the failure mechanism: mirror, mist or hackle-like and crack branching[15-17]. The rupture pattern and its boundaries were used to identify the size of the initial flaw[23-24], assess toughness[18,19], strength[19-22], or residual surface stress[21,24].

Single fiber tensile properties were measured on a micro-tensile testing machine at a constant loading rate of 0.5%.min[-1] using a gauge length of 25 mm. In order to obtain a good statistical analysis, 40 to 50 samples were tested for each type of fiber. The fiber diameter was measured by laser diffraction prior to testing, and was observed post mortem by SEM.

SEM Observations

SEM observations of fracture surfaces are aimed at identifying the crack origin, measuring the area of typical flaw features and checking the diameter of the fiber. Among the several hundreds of rupture surfaces that were observed by SEM, seven typical families of characteristic flaws were identified: five are originating from surface flaws and two from internal flaws. These characteristic flaws are shown in Figure 5. The A-type flaws have a well-known penny shape contour, the B flaws are small particles present at the surface of the fibers, the C flaws are small surface damages that may result from manipulation or contact between fibers, and the D flaws are chemical composition heterogeneities[17]. In the case of E-type flaws, though rupture clearly originates from the surface, the flaws cannot be observed by SEM because of their particularly small size or the poor image contrast. Concerning the rupture induced by internal flaws, the F flaws are internal inclusions[25] or voids and G flaws are typical penny shaped cracks (similar to A). These latter flaws were only observed on Hi-Nicalon fibers. A statistical analysis of the flaws responsible for the fiber ruptures is summarized in Figure 6. When plotting, for each fiber, the proportion of ruptures originating from internal and surface flaws, the behavior of the Hi-Nicalon fiber is clearly different from that of the other fibers: almost all their observed ruptures originate from internal flaws. This result is in accordance with G. E. Youngblood et al.[25] observations stating that all of the Hi-Nicalon fiber ruptures originate from internal flaws, though J. Hurst[26] only observed 60% of ruptures induced by internal flaws. As a general trend, the proportion of Nicalon fiber ruptures induced by internal flaws is higher than that of Tyranno fibers, except for the Lox-M fiber.

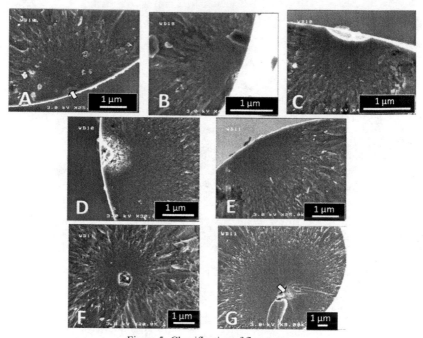

Figure 5: Classification of flaw pattern
Surface flaws: (A) penny shaped crack, (B) particle, (C) surface damage, (D) chemical heterogeneity, (E) no observable trace by SEM,
Internal flaws: (F) internal inclusion, (G) penny shaped crack

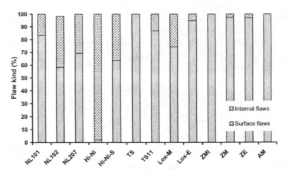

Figure 6: ratio between surface and internal flaws responsible for ruptures for the various fibers

During tensile tests, different stages of crack propagation are usually observed, corresponding to four concentric domains of the fracture pattern[16] (Figure 7). Close to the initial flaw, a smooth mirror region corresponds to an increase of the crack velocity that reaches its maximal value at the edge of this domain. In the mist domain surrounding the mirror domain, the crack dissipates energy by generating micro cracking. As the crack propagates, more energy is available and larger cracks nucleate in a domain referred to as hackle. The final stage corresponds to macroscopic cracks propagating through the entire section of the fiber.

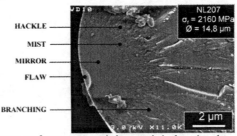

Figure 7: Typical fracture surface pattern and characteristic domains observed on SiC fibers.

Strength and Toughness

From experimental work carried out by many authors on mono or polycrystalline brittle materials, it can be inferred that the strength of the fiber is proportional to the inverse square root of the radius of the mirror domain[16, 27, 28]. Such a relationship was determined for the various fibers and is illustrated in the case of NL207 fiber (figure 8). The experimental points were fitted by a straight line assuming that the residual stress resulting from synthesis can be neglected (straight line drawn through zero). The value of the deduced mirror constant (2.37 MPa.m$^{1/2}$) is close to the Am values reported by L. C. Sawyer (2 MPa.m$^{1/2}$)[17] or A. J. Eckel (2.51 MPa.m$^{1/2}$)[20] on the same fiber.

The same relationship can be established using the radius of hackle or branching domains:

$$\sigma_r = \frac{A_m}{\sqrt{r_m}} \qquad \sigma_r = \frac{A_h}{\sqrt{r_h}} \qquad \sigma_r = \frac{A_b}{\sqrt{r_b}}$$

(2)

where A_m, A_h and A_b are constants corresponding respectively to mirror, hackle and branching domains and r_m, r_h and r_b their respective radii.

Though Am is almost exclusively used by the authors, the A_h or A_b constants can provide the same results and are easier to measure by fractographic examination. Indeed, the smaller size of the mirror region and the difficulty to accurately determine its boundary with the mist region are a source of error, which may explain the discrepancies found in literature. We found A_m values ranging from 1.7 to 3.0, A_h in the range 1.9 - 3.6 and A_b in the range 2.5 - 4.7 MPa.m$^{1/2}$. A_m constants are similar to typical values measured on glass materials.

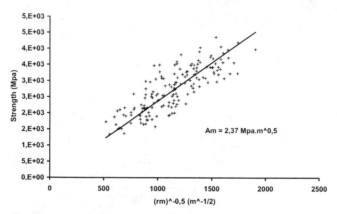

Figure 8: Tensile strength plotted as a function of the inverse square root of the radius of the mirror domain

The fiber fracture behavior depends on its toughness which is quantified by the critical stress intensity factor (K_{IC}). This factor was calculated for all fibers by use of the Griffith equation:

$$K_{IC} = Y\sigma_r\sqrt{a_c} \qquad (3)$$

where Y is a geometrical factor (for a penny shaped flaw: $Y = 2/\sqrt{\pi}$), σ_r is the fiber strength and a_c is the flaw size. This calculation was only carried out in the case of penny shaped flaws, the surface of other flaws being difficult to measure. Toughness values of all the SiC fibers of first and second generation, except for Hi-Nicalon, typically range from 1.0 to 1.3. Third generation fibers that are nearly-stoichiometric, have higher toughness but always lower than 2 MPa.m$^{1/2}$.

Combining equations (2) and (3) leads to the definition of a proportional coefficient between the mirror constant and the toughness according to:

$$\frac{2A_m}{\sqrt{\pi}K_{IC}} = \sqrt{\frac{r_m}{a_c}} \qquad (4)$$

CONCLUSION

All these characterizations reveal a strong correlation between the microstructure of fibers and their oxygen content. Oxygen is introduced into the fibers during the cross-linking reaction of the polymer precursor. This stage has thus a strong influence on the amount of amorphous Si-C-O formed in the fiber, which is almost proportional to the oxygen content. In particular, the amount and size of β-SiC crystallites are clearly correlated to the oxygen content of the fiber: the higher the oxygen content, the less the amount of β-SiC crystallites, the smaller the β-SiC crystallites. This behavior is in accordance with the location of the composition of the fibers in the isothermal section of the Si-C-O phase diagram. Both measurements by chemical analysis and NMR of the C/Si ratio in the Si-C-O continuum of the various fibers are in good agreements. This result tends to confirm that the "free carbon" phase, embedded into the continuum is not bonded to the amorphous $SiC_xO_{2(1-x)}$ glass nor the β-SiC grains. The amount of free carbon measured by both chemical analysis and NMR are in good accordance. They show that free carbon content of Tyranno fibers is much higher than that of Nicalon fibers, as the result of the type of polymer used as precursor.

Single fiber tensile properties were measured on a micro-tensile testing machine and their fracture surface observed by SEM. Looking at the proportion of ruptures originating from internal or surface flaws, reveals a clearly different behavior of the Hi-Nicalon fiber as almost all their observed ruptures originate from internal flaws. As a general trend, the proportion of Nicalon fiber ruptures induced by internal flaws is higher than that of Tyranno fiber.

ACKNOWLEDGEMENT

Financial support from the French programme ARCOCE funded by FUI and Aquitaine Region is gratefully acknowledged.

REFERENCES

[1] J. Hayashi, K. Okamura, S. Yajima, Structural analysis in continuous silicon carbide fiber of high tensile strength, *Chemistry Letters,* 1209-12 (1975)

[2] H. Ichikawa, Development of high performance SiC fibers derived from polycarbosilane using electron beam irradiation curing-a review, *J. Ceram. Soc. Jap.*, **114** 455-60 (2006)

[3] C. Laffon, A.M. Flank, P. Lagarde, M. Laridjani, Study of Nicalon-based ceramic fibers and powders by EXAFS spectrometry, X-ray diffractometry and some additional methods, *J. of Mater. Sci.,* **24**, 1503-12 (1989).

[4] L. Porte, A. Sartre, Evidence for a silicon oxycarbide phase in the Nicalon silicon carbide fibre, *J. of Mater. Sci.,* **24**, 271-75 (1989).

[5] P. Le Coustumer, M. Monthioux, A. Oberlin, Understanding Nicalon Fiber, *J. of the Eu. Ceram. Soc.,* **11**, 95-103 (1993).

[6] N. Hochet, M. H. Berger, A. R. Bunsell, Microstructural evolution of the latest generation of small-diameter SiC-based fibers tested at high temperatures, *J. of Microscopy*, **185**, 243-58 (1997).

[7] A.R. Bunsell, A review of the development of three generations of small diameter silicon carbide fibres, *J.Mater.Sci* **41**, 823-39 (2006).

[8] G. Chollon, PhD thesis, Fibres céramiques à base de carbure de silicium et à faible taux d'oxygène, (1995)

[9] R. Bodet, N. Jia, R.E. Tressler, Microstructural Instability and the Resultant Strength of Si-C-O (Nicalon) and Si-N-C-O (HPZ) Fibers, *J. Am. Ceram. Soc.*, **16**, 653-664 (1996).

[10] P. Kroll, Modelling and simulation of amorphous silicon oxycarbide, *J. Mater. Chem.*, **13**, 1657-1668 (2003).

[11] P. Kroll, Modeling the free carbon phase in amorphous silicon oxycarbide, *Journal of Non-Crystalline Solids,* **35,** 1121-26 (2005).

[12] P. Scardi, M. Leoni, R. Delhez, Line broadening analysis using integral breadth methods: a critical review, *Journal of applied crystallography*, **37**, 381-90, (2004)

[13] Th.H. Keijser, E.J. Mittemeijer, H.C.F. Rozendaal, The determination of crystallite-size and lattice-strain parameters in conjunction with the profile refinement method for the determination of crystal structures, *Journal of Applied Crystallography*, **16**, 309-16, (1983)

[14] S. M. Dong, G. Chollon, C. Labrugère, M. Lahaye, A. Guette, J. L. Bruneel, M. Couzi, R. Naslain, D. L. Jiang, Characterization of nearly stoichiometric SiC ceramic fibres, *Journal of Materials Science*, **36**, p. 2371-81 (2001)

[15] J. W. Johnson, D. G. Holloway, On the shape and size of the fracture zones on glass fracture surfaces, *Philosophical Magazine*, **14**, p. 731-43 (1966)

[16] J. J. Mecholsky, S. W. Freiman, R. W. Rice, Fractographic analysis of ceramics, *American Society for Testing and Materials*, p. 363-79, (1978)

[17] L. C. Sawyer, M. Jamieson, D. Brikowski, M. H. Haider, R. T. Chen, Strength, structure, and fracture properties of ceramic fibers produced from polymeric precursors: I, base-line studies, *Journal of the American Ceramic Society*, **70**, 11, p. 798-810 (1987)

[18] I. J. Davies, T. Ishikawa, Estimation of the fracture toughness of Tyranno Si-Ti-C-O fibres from flaw size and "fracture mirror" data measured in situ a 3-D woven SiC/SiC composite, *International Journal of Materials and Product Technology*, **16**, p. 189-96 (2001)

[19] K. Morishita, S. Ochiai, H. Okuda, T. Inchikawa, M. Sato, Fracture toughness of a crystalline silicon carbide fiber (Tyranno-SA3®), *Journal of the American Ceramic Society*, **89**, 8, p. 2571-76 (2006)

[20] A. J. Eckel, R. C. Bradt, Strength distribution of reinforcing fibers in a Nicalon fiber/chemically vapor infiltrated silicon carbide matrix composites, *Journal of the American Ceramic Society*, **72**, 3, p. 455-58 (1989)

[21] B. J. Norman, A. C. Jaras, J. Ashall, Measurement of silicon carbide fibre strength in composites from studies of their fracture surfaces, British Ceramic Transactions, **92**, 2, p. 62-66 (1993)

[22] I. J. Davies, T. Ishikawa, M. Shibuya, T. Hirokawa, Fibre strength parameters measured in situ for ceramic matrix composites tested at elevated temperature in vacuum and air, *Composites Science and Technology*, **59**, 6, p. 801-11 (1999)

[23] M. D. Thouless, O. Sbaizero, L. S. Sigl, A. G. Evans, Effect of interface mechanical properties on pullout in a SiC-fiber-reinforced lithium aluminium silicate glass-ceramic, *Journal of the American Ceramic Society*, **72**, 4, p. 525-32 (1989)

[24] S. T. Taylor, Y. T. Zhu, W. R. Blumenthal, M. G. Stout, D. P. Butt, T. C. Lowe, New perspective on the fracture of Nicalon fibers, *Fatigue and Fracture Mechanics*, **29**, p. 1-11 (1998)

[25] G. E. Youngblood, C. Lewinsohn, R. H. Jones, A. Kohyama, Tensile strength and fracture surface characterization of Hi-Nicalon SiC fibers, *Journal of Nuclear Materials*, **289**, p. 1-9 (2001)

[26] J. Hurst, H.-M. Yun, D. Gorican, *Advances in ceramic-matrix composites III, Editors N. P. Bansal, J. P. Singh*, CT 74, p. 3 (1996)

[27] W. C. Levengood, Effect of origin flaw characteristics on glass strength, *Journal of Applied Physics*, **29**, p. 820-27 (1958)

[28] L. C. Sawyer, R. Arons, F. Haimbach, M. Jaffe, K. D. Rappaport, Characterization on Nicalon: strength, structure, and fractography, *Ceramic Engineering and Surface Proceedings*, **6**, 7-8, p.567-75 (1985)

THE CONTROL OF INTERPHASES IN CARBON AND CERAMIC MATRIX COMPOSITES

Gérard Vignoles, René Pailler, Francis Teyssandier
Laboratory for Thermostructural Composites, University Bordeaux 1, Pessac, France.

ABSTRACT:
    Toughness and damage-tolerance of ceramic matrix composites (CMC) are achieved by engineering the fiber/matrix interface in order that a crack initiated in the brittle matrix does not propagate in a catastrophic manner through the reinforcing fibers. Since the 1970s, many solutions ranging from simple interface response to addition of a monolayer or multilayered interphase have been imagined and developed. Among them the pyrocarbon (PyC) interphase may be considered as a reference. However, its performance is highly dependent on its nanotexture that has to be accurately tailored through its fabrication process: chemical vapor deposition or infiltration (CVD/CVI). Control of pyrocarbon microstructure by CVI is a key issue in the processing of PyC interphases in CMC as well as of matrices in high-performance carbon/carbon composites. The gas-phase chemistry plays a key role in the various nanotextural transitions. The relation between nanotexture and processing conditions has been elucidated by joint experimental and modeling approaches.

KEYWORDS: fiber/matrix interphases, CVD of Pyrocarbon.

INTRODUCTION

    Although man knows empirically for thousands of years how to improve the behavior of brittle materials such as straw-reinforced mud bricks, or more recently to improve the mechanical properties of materials by use of a composite structure (Mongol bow), the controlled elaboration and further understanding and improvement of ceramic matrix materials are only 40 to 50 years old. Carbon/carbon (C/C)[1] and ceramic matrix composites (CMC)[2] are nowadays used to produce at an industrial scale, parts of rockets or aircrafts: engines nozzle, divergent skirt, heat shields for atmospheric reentry, brake discs... These materials are furthermore aimed at being used in civil aeronautics to replace an increasing number of metallic parts: aircraft engine plug or exhaust, compressor blades...).

    In order to allow these composite materials to become reliable and reproducible industrial products, several main issues had to be overcome. Among them, the first difficulty was to find a way to confer a reasonably tough and damageable behavior to a composite material that comprises exclusively brittle constituents. CMC differ significantly from other types of composites as the failure strain of the matrix is lower than that of the fiber. For that reason, they are called inverse composite as compared to metal/metal composites (MMC) or organic matrix composites (OMC): under tensile load in the direction of the fibers, the matrix fails first. Back to the mid 1960s Cook & Gordon[3] stated that "Inside a brittle solid, if a plane of weakness or potential cleavage is present and is roughly normal to the plane of the original crack, this interface may break and produce a secondary crack in such a manner as to interfere with the progress of the primary crack."

    It is now well established that the mechanical behavior of such materials is critically dependent on the bonding between fiber and matrix and it has been further demonstrated that this bonding is also critical for oxidation resistance of the composite at high temperature in oxidative environments. This property gives the specific added value to fibers embedded in a matrix: the whole is greater than the sum of the parts. However, in order to be efficient, the bonding has to be adjusted at the "proper" value: if the fiber/matrix (F/M) bonding is of the same order of magnitude

11

than that of the matrix, it is called strong and the composite is brittle; in contrast, if the F/M bonding is weaker, the composite becomes non-brittle. Of course, if it is too weak, then the matrix plays no role in the composite. In other words, the strength of the bonding must be sufficiently high to allow an efficient transfer of the load applied to the matrix, towards the reinforcing fiber, but sufficiently weak to prevent the cracks generated in the matrix to propagate through the fiber and induce the ruin of the composite.[4,5]

After a short review of the various interphases that can be used in SiC/SiC composite materials, the present paper will focus on the understanding of the chemical processes leading to the formation of the large variety of pyrocarbon nanotextures (or microstructures). The control of the pyrocarbon microstructure is a main issue of the CVD/CVI process both for SiC/SiC (interphase) and Carbon/Carbon composites (matrix). The influence of the process on the structure and properties of the deposited pyrocarbon are detailed. The influence of the nature of the precursor is discussed.

INTERPHASES IN SiC/SiC

The F/M bonding is indeed not a surface or interface which could be understood as a pure 2D domain. As the result of physicochemical processes (chemical reaction, diffusion) that takes place at high temperature, it is usually composed of one or many thin phases. These phases may form a third body between fiber and matrix or extend themselves inside the fiber. A huge amount of literature has been devoted to the characterization and understanding of a domain the thickness of which typically ranges from 10 to 100 nm, and has accordingly to be investigated at nanometer scale. Readers are referred to a review[4] on the subject and we only mention here the main milestones.

The simplest bonding approach is to promote and control the formation of a weak interphase during the high temperature step of the processing. This approach has been used for the development of glass-ceramic matrix composites[6]. First generation Si-C-O fibers that contain excess carbon and oxygen form a carbon rich fiber/matrix interfacial layer when incorporated into glass-ceramic matrices at elevated temperatures, which is responsible for the high toughness observed.

Another approach is based on the deposition of a thin coating on the fiber prior to its densification stage to form the matrix. As the additional phase may be designed as a single layer or a multilayered coating, it offers much more flexibility. A large variety of phases or combination of phases has been applied to C/SiC and SiC/SiC composites. The most commonly used are materials exhibiting a layered crystal structure: PyC, hex-BN, BCN[7]. However, PyC is prone to oxidation above 500°C, hex-BN is moisture sensitive when prepared at low temperature and BCN is difficult to prepare at a desired composition. Other phases exhibiting a layered crystal structure and showing furthermore a good oxidation resistance belong to the MAX-phase family. Their interesting combination of properties is traceable to their layered structure, including a combination of mostly metallic strong M-X bonds, together with relatively weak M-A bonds allowing shear cleavage. The ternary most-studied as interphase to date is Ti_3SiC_2[8,9]. However, as a CVD deposit, is has been up to now unable to deflect cracks, since its layered structure has not been obtained parallel to the fiber surface. The concept of intrinsic nano-layered material has been extended to stacks of layers made of different materials[4,5]. Such a concept allows a wide choice of materials that can be tailored to deflect cracks and/or improve oxidation resistance behavior, either by use of oxidation resistant materials or by formation of healing materials. Though a large variety of different layers can be associated, thus generating a wide field of new fiber/matrix interphases, such nanometer-thick multilayered structures are difficult and expensive to control at an industrial scale. Several examples are encountered in the literature, many of them having been summarized in reference 4.

PYROCARBONS

Among all the interphases described above, pyrocarbon or pyrolytic carbon[10] is still, in spite of its poor oxidation resistance, a reference material. It is obtained by dehydrogenation at high temperature of a gaseous hydrocarbon. For CMC or C/C fabrication purposes, it is usually grown by CVD or CVI process on a hot fibrous substrate. The structure of pyrocarbon is similar to that of graphite but includes disorder (figure 1): graphene layers have limited extent and may include C5 or C7 arrangements responsible for some waviness; they may furthermore be stacked with rotational disorder (turbostratic graphite) and contain screw dislocations.

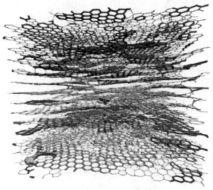

Figure 1: Atomistic model of pyrocarbon obtained by Image Guided Atomistic Reconstruction (IGAR) method[11]

The versatility of disorder features leads to various nanotextural forms of pyrocarbon[12-16], ranging from nearly isotropic to highly anisotropic (*i.e.* close to graphite structure). Various denominations are encountered according to distinct scientific and technological teams. The characterization of PyC nanotexture and its relation to processing conditions have been the object of investigations since more than 50 years.

A key issue in the CVD or CVI of pyrocarbons is the control of the nanotexture of the deposit during processing. In the case of SiC/SiC interphase, the graphene layers have i) to be oriented parallel to the fiber surface, and ii) to form a bond with the fiber surface, which must be strong enough in order that the debonding induced by crack deflection occurs at the interphase/matrix interface. In the case of carbon/carbon composites, the basic questions to be addressed are i) how to obtain a uniform dense matrix in the preform in a reasonable time, ii) how to control the homogeneity of the microstructure through the whole thickness of the part to be infiltrated. For both applications, another question is which carbon precursor to choose among the numerous potential hydrocarbon molecules. Though easy to implement, the CVD/CVI process is extremely complex as it involves numerous homogeneous as well as heterogeneous chemical pathways, both in the reactor and in the preform, fluid flow, heat and mass transfer. Accordingly, the basic need of understanding in order to optimize the fabricated material is high. CVD and CVI pyrocarbons have been the object of studies since many decades and is still a topic of actuality, because of the complexity of the involved chemistry, and of its subtle entanglement with transport phenomena.

Starting in the early 1960s, the main improvements concerning CVD/CVI of matrices in C/C composites have been recently reviewed by R.J. Diefendorf[17]. The paper discusses the merits of the various types of CVI processes: isothermal or with temperature-gradient, with or without forced or pulsed flow. It also addresses the thermodynamics of the carbon/hydrogen system. Soon[18-19] he had recognized that, starting from methane, temperature and pressure were critical parameters influencing the organization of the deposited carbon. For example, fig.2 shows[18] – in the case of CVI from methane – the distinction between isotropic (sooty) carbons and anisotropic (laminar) pyrocarbons, the first ones being the result to some extent of homogeneous nucleation, while the second ones were more linked to heterogeneous processes. However, this description is highly dependent upon reactor geometry and residence time and has accordingly been subject to strong controversies.

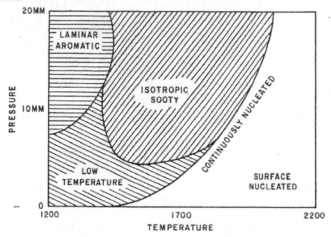

Figure 2: Fields of microstructure for pyrolytic graphite deposited from methane according to ref. 18

Some time later, several kinds of laminar pyrocarbons were distinguished.[20,21] Among them, two varieties have been identified, and have been called "Rough Laminar" (RL) and "Smooth Laminar" (SL) because of their appearances when imaged by Polarized Light Optical Microscopy (PLOM), as described by Diefendorf & Tokarsky[22]: when rotating the cross-polarized light, the orthogonal branches of the Maltese cross show a 'rough' extinction in the case of Rough Laminar variety, though extinction is 'smooth' in the case of smooth laminar variety. Rough and smooth PyCs differ by their degree of structural anisotropy, and have distinct properties. For instance, the RL form is graphitizable by a high-temperature post-treatment[13,14]. Much more recently, a new form of highly anisotropic PyC, distinct from RL, has been identified and termed "Regenerative Laminar".[23] Raman analysis under polarized light has played a crucial role in separating these forms from each other.[24] Confusion between RL and ReL has been common until very recently. There are still classifications which are only based on the optical or Small-Area Electron Diffraction (SAED) anisotropy[25,26] and do not provide a distinction between RL and ReL.

The current state of pyrocarbon denominations, together with a detailed description based on PLOM imaging, extinction angle, High Resolution Transmission Electron Microscopy - Lattice Fringe (HRTEM - LF) imaging, SAED Opening Angle, and Polarized Raman signals, is summarized in Table 1.

Table 1. Classification and description of pyrocarbons, in relation with processing parameters.

	Rough Laminar (RL)	Smooth Laminar (SL)	Regenerative Laminar (ReL) (was formerly confounded with RL)	Dark Laminar (DL)
Recommended name[15]				
PLOM visualization on fibers and extinction angle (Ae°)	[15-23]	[10-14]	[15-23]	[5-10]
HRTEM LF visualization		5nm	5nm	
SAED figure and Opening Angle OA°	[20-40]	[40-90]	[20-40]	[80-90]
Raman anisotropy ratio R_A (-)	[5.5-10]	[4-5.5]	[5.5-10]	[4-4.5]
Raman band width $FWHM_D$ (cm^{-1})	[80-140]	[80-200]	[140-200]	[80-200]
Graphitizable ?	Yes	No	Yes	No
ICT Karlsruhe classification[25-26]	High Textured (HT)	Medium Textured (MT)	High Textured (HT)	Low Textured (LT)
F. Langlais et al. Classification[58]	Columnar Laminar (CL)	Weakly Anisotropic Laminar (WAL)	Highly Anisotropic Laminar (HAL)	Quasi-Isotropic Laminar (QIL)
Processing conditions	Increasing T ; increasing t_s ; decreasing S_V; increasing hydrocarbon source gas reactivity →			

PYROCARBON NANOTEXTURE: PARAMETER INFLUENCE AND MECHANISM

It has been shown[18-21, 27-29] that the most influential processing parameters are, in addition to temperature T and pressure P, and initial gas-phase composition: residence time t_s and deposition space volume/deposition surface area ratio, S_V[30]. One of the most confusing points is that CVD observations and models are not readily transposable into CVI models, for at least three reasons. First, because of the high S_V values, heterogeneous chemistry plays a much more important role in CVI in contrast to CVD[30]. For instance, if saturation adsorption may occur in CVD, this is much less possible in CVI, at least far from the porous medium surface. Also, some pyrolysis sub-mechanisms which exist in the gas phase may be completely overridden by heterogeneous ones, leading to a very different apparent behavior of the gas phase. Second, depletion effects are to be expected in narrow pores, due to transport limitations. This latter point has been thoroughly investigated in numerous modeling works, dealing either with isothermal-isobaric CVI (I-CVI),[31-34] forced-CVI (F-CVI),[35] thermal-gradient-CVI (TG-CVI) -- either with microwave or radio-frequency heating (see a review in Ref. 36). Third, the depletion effects that are to be expected outside the fibrous preform may be very pronounced in CVI cases that correspond to very high S_V values.[37] The preform is much more reactive than a plane substrate, so that the diffusion boundary layer which surrounds it is larger, and may easily reach the outer walls.[38] Taking these facts into account is only feasible in a global modeling frame where both the preform and the surrounding free-medium are simultaneously considered.[39]

Many experimental studies have shown in the past the importance of processing parameters on the PyC deposition rates and nanotexture, in various physico-chemical conditions and reactor configurations, either in CVD (plain substrate, low S_V), or in CVI (porous substrate, high S_V).[40-58] Most of them have tried to identify some "ultimate precursor" of PyC, either light, aliphatic species, or heavy aromatic compounds such as Polycyclic Aromatic Hydrocarbons (PAHs) and polyynes. It has been proved, from mass spectroscopy,[49,53,29] gas chromatography,[54,55] and FT-IR[56,57,58] measurements of the gas-phase composition, that the hydrocarbon pyrolysis follows a long chain of homogeneous reactions, in a so-called "maturation" process : i) precursor decomposition, ii) recombination of the first products into other species among which unsaturated species and Resonance-Stabilized Free Radical species (RSFRs), iii) growth of heavier molecules with a varying degree of unsaturation or aromaticity, among which PAHs.

One the key issues of pyrocarbon nanotexture control is the identification of the most important species and sub-mechanisms that may be associated to a given nanotexture. R.J. Diefendorf[59] contributed to the pioneering work devoted to the understanding of the various mechanisms involved in the deposition of pyrolytic carbon. He was among the first to identify the influence of gas phase mechanisms on the switching between distinct growth processes: either growth from the edges or deposition on the surface of basal planes, according to the nature of the molecules formed in the gas phase. In the case of heavy precursors, a condensation-like mechanism has been proposed,[59,60] while for light species, a lateral growth mechanism, close to a radical-based polymerization reaction scheme, are more likely.[59,61] It has to be expected that the overall reaction rate dependence on temperature should vary with temperature (since the activation energies may be different), on the total pressure and residence time (since the relative amounts of various key species may vary), and on S_V (since the heterogeneous to homogeneous reaction ratio varies). Indeed, all these parameters may toggle the system between various dominant mechanisms.

Actually, the experimental data, as collected from propane and methane pyrolysis and subsequent CVD, were puzzling. Fig. 3 summarizes the numerous textural transitions that have been observed in methane[62,63] and propane[55-58,64-66] pyrolysis.

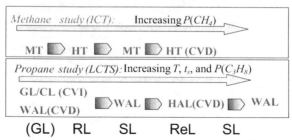

Figure 3: Observed textural transitions in methane studies at ICT[62,63] and propane studies at LCTS.[55-58,64-66] Notations are explained in Table 1.

In both cases, when processing parameters acting directly on gas-phase maturation are varied, there is a non-monotonous evolution of the anisotropy degree. The only point is that sometimes the data from distinct authors were in apparent contradiction. Guellali[62] reported in model capillaries (moderate S_V, close to CVD conditions) a "(low-)high-low-high" anisotropy sequence while in CVD Féron[56,57] reported a "low-high-low" sequence. At first, no attention was paid to distinguish between RL and ReL (both being highly anisotropic). Later on, it was understood that the second highly anisotropic PyC appearing in the propane study was ReL as opposed to RL at low residence times and high S_V ratios. Things went thus clearer when LCTS managed to produce highly anisotropic PyC at extremely short residence times: it was RL and not ReL. The anisotropy evolution sequences could then be matched as depicted in figure 3.

PYROCARBON DEPOSITION: MODEL BUILDING

From a chemical kinetic point of view, the modeling efforts concerning PyC deposition have been devoted primarily to the study of the maturation phenomenon, that is, to hydrocarbon pyrolysis. The gas-phase reaction steps have been successfully modeled using comprehensive[67-71] or semi-detailed[72] mechanisms. They have been completed by the incorporation of heterogeneous reactions,[73-74] but with less detail due to the lack of knowledge concerning these reactions.

In the case of propane decomposition at $P = 2$ kPa, $T = 900°C$-$1200°C$, and moderate to large residence times, in a CVD tubular reactor, a kinetic modeling study based on a detailed gas-phase mechanism has confirmed the maturation phenomenon and its importance on the transition from SL to ReL (mistakenly called RL in the papers) when residence time and/or temperature increases.[67,68] It appeared that only the heaviest species in the detailed model could account for ReL pyrocarbon growth, while many lighter ones were related to SL. A very detailed study has been devoted later on to the growth of SL pyrocarbon,[69-71,73,74] and one of the results is that SL PyC growth is equally due to the presence of CH_3 radicals and C2 molecules like acetylene and ethylene.

A CVD simplified model has been deduced from the detailed pyrolysis study[68] in order to define a computationally tractable chemical model to be included in the full scale modeling of the reactor. Its simplified homogeneous chemical path takes into account two groups of key species: aliphatic species (mainly CH_4, C_2H_6, C_2H_2 C_6H_6 according to the initial hydrocarbon molecule) that contribute to the growth from the edges of carbon basal planes, and PAHs species that contribute to the growth of the surface of carbon basal planes through a condensation-like mechanism. Note that the kinetic constants associated with the formation of intermediate aliphatic species CH_4 and C_2H_6 or C_2H_2 or C_6H_6 takes implicitly into account the whole detailed gas-phase mechanism including intermediate species. For that purpose, experimental data have been used for the identification of some rate constants and activation energies. It is worth noting that this CVD model accounts for the last ReL→SL transition at large residence times, due to PAH depletion. For the full-scale modeling of the CVI reactor including the simplified chemical model, [38,75] the porous medium model was such that its effective transport coefficients were known precisely enough during infiltration. Thus, one-dimensional CVI computations have been performed in order to reproduce the deposit thickness profiles. The model manages to reproduce successfully either CVD or CVI experiments using propane. This model has also been injected into a coupled solver featuring both free-medium and porous medium, and allowed to understand how ReL PyC appears by maturation then disappears by depletion due to the large deposition rate of PAHs. Deposition experiments in carbon foams, with moderate S_V ratios, were successfully modeled.[76]

In the case of PyC deposition from methane, the detailed models of pyrolysis[77] show that the characteristic induction times for maturation are considerably higher than those for propane, due to the exceptional stability of the methane molecule. This arises from the high activation energy for C-

H bond breaking, which is crucial in the first step of methane decomposition chemistry. Additionally, this first step is a "third-body enhanced" reaction:

$$CH_4 + M \rightarrow H + CH_3 + M$$

which gives an important role either to total pressure or to S_V (M being any gaseous species or the surface, respectively). This special feature of CH_4 has been successfully used to undertake inside-out isothermal CVI.[78] Once the pyrolysis has begun, attention is paid mostly to acetylene and benzene as key intermediates for the growth of PyC at low residence times, and PAHs for large residence times. This is in complete consistency with the propane-based chemical scheme.

More recent simulation studies[79,80] with parameter identification based on CVD[51] and CVI[54,30-31] experiments tend to confirm the idea that groups of molecules C_2H_2, C_6H_6 and PAHs are also key species in the case of deposition from methane. The main difference is now that CH_4 should be considered both as the initial hydrocarbon molecule for the mechanism as well as an intermediate species; the formal scheme presented for propane does not have to change, only the H_2 production stoichiometry has to be revised.

There remains to explain why highly anisotropic PyC is deposited at very short residence time and high S_V ratios, and this is more complicated. Low molecular weights molecules are obviously involved in the process and there is strong evidence that key species have to be early products of pyrolysis (first- or second-generation products, like the CH_3 radical, C_2H_4, etc ...).

It has been argued that HT deposition from methane (probably RL) results mainly from aliphatic species, since, at the considered experimental conditions (higher P and T), homogeneous nucleation of soot overrides the formation of HT PyC. To explain this, a "particle-filler" model has been proposed,[79] asserting that the degree of anisotropy is susceptible to increase when a particular C_2H_2/C_6H_6 ratio is reached, this phenomenon taking place at short residence times. This model states that very anisotropic pyrocarbon may be deposited by lateral growth of graphene planes, for which the lowest quantity of defaults (e.g. C5 or C7-rings, helicoidal structures) is attained when C_2H_2 acts as a "filler" between C_6H_6 "particles". However, it is difficult to imagine the precise bimolecular or multi-molecular mechanism through which C_2H_2 and C_6H_6 would yield perfectly matching additions. Numerical results from the detailed mechanism[80] also give two pieces of information on the propane and methane cases: first, C_2H_2 and C_6H_6 have a very similar variation, even though C_6H_6 is formed later; second, the C_6H_6/C_2H_2 ratios correlated to RL deposition are neatly different in the two cases. There is no obvious explanation for this latter fact with the "particle-filler" model.

On the other hand, there have been tentative explanations based on the topology of carbon edge growth from molecules and radicals containing even or odd numbers of carbon atoms. In the case of propane, it has been proposed[80] that the very early stages of decomposition (or no decomposition at all) favor the relative abundance of C3-containing species, while later stages of maturation increase the C2-containing species amounts; in his review[17], R. J. Diefendorf suggests that C2 species are capable of creating C5 cycles, and consequently local bending of the graphitic sheets.[81] This fact is also consistent with the RL→SL transition in the case of methane – indeed, detailed modeling yields CH_3 as a first radical, and C_2H_2 appears much later during pyrolysis; if CH_3 is dominant in the gas-phase, then hexagon growth is preferred and the anisotropy is large, while if C_2H_2 dominates, then faulted cycles are more abundant and the anisotropy decreases.

Thus, a global model of PyC deposition should take into account the gas-phase maturation (i.e. the onset of various species by bond breaking and recombination), and deposition mechanisms originating in various precursors, since the non-monotonous anisotropy transitions clearly suggest a variety of heterogeneous reactions. Concerning homogeneous reactions, it seems that propane and methane show a similar behavior, at the noticeable exception of the initiation step, which is much slower for methane.

First, highly anisotropic PyC deposited at long residence time (ReL) is clearly related to high molecular weight species such as PAHs as established both from experimental[55-58] and computational[68,75] facts. In contrast, the intermediate MT (SL) and low-residence time HT (RL) forms of PyC seem to be correlated to species with lower molecular weights,[69-71,73,74] the presence of C_2H_2 (and C_6H_6) inducing a loss of anisotropy and triggering the RL → SL transition.

The qualitative schemes of the propane-based and methane-based mechanisms for PyC deposition are summarized in figure 4. A companion paper[82] discusses its identification and implementation, with a successful validation with respect to experimental data in the case of propane pyrolysis.

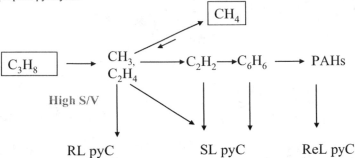

Figure 4: simplified homogeneous chemical path deduced from the detailed mechanism.

CONCLUSION

This document has summarized some key aspects in CMC and C/C composite development to which the contribution of R. J. Diefendorf has been of large importance for the scientific and industrial community. His ideas have been starting points and guidelines for many studies which have led to the current state of knowledge on interphases, and principally pyrocarbon ones.

Though pyrocarbon, almost exclusively composed of carbon (and less than 1at% hydrogen), can be seen as the simplest material for matrix or interphase purposes, it shows many varieties of textures that have to be controlled. Indeed, the ability of PyC to graphitize or to show some desired properties requires specific nanotexture that must be tailored by the manufacturing process. Materials science has taught us the exclusive relationship between a process and the properties of the produced material. Apart from the usual parameters controlling the process (T, P, initial gas-phase composition), two parameters are of critical importance in the growth of PyC from chemical vapor deposition: residence time and deposition space volume/deposition surface area ratio. Both are related to homogeneous gas phase reactions during their transport towards the surface to be coated and their competition with heterogeneous reactions responsible for the growth itself. Accordingly, the understanding and control of the desired PyC properties requires in this case an extended knowledge ranging from solid state chemistry to gas phase reactivity by relying on chemical engineering.

Even though the control of the pyrocarbon nanotexture, and thus of the composite toughness is now at hand, there is still work to perform on the subject, both from the industrial side – process optimization, control of manufacture – and from the basic science side – for instance, heterogeneous chemistry modeling is still in its infancy when one deals with such a fascinating material as pyrocarbon.

ACKNOWLEDGEMENT
G. L. Vignoles acknowledges ANR for grant 2010-BLAN-0929 to his current research on pyrocarbon matrices nanotexture.

REFERENCES
[1]G. Savage, Carbon-Carbon Composites, Chapman & Hall, (1993)

[2]R. Naslain, Fibre-matrix interphases and interfaces in ceramic matrix composites processed by CVI, *Composite Interfaces,* **3**, p253-286 (1993)

[3]J. Cook, J.E. Gordon, C. C. Evans and D. M. Marsh, A mechanism for the control of crack propagation in all-brittle systems, *Procs. Roy. Soc. A*, **282** p.508-20 (1964)

[4]R. Naslain, R. Pailler, J. Lamon, Single and multilayered interphases in SiC/SiC composites exposed to severe environmental conditions: an overview, *Int. J. Appl. Ceram. Technol.*, **7**, p.263-75 (2010)

[5]R.P. Boisvert, R.K. Hutter, R.J. Diefendorf, Interface manipulation in ceramic matrix composites for improved mechanical performance, in Procs 4th Japan-US conference on composite materials, Washington, June 27-29, 1993, p. 789-98 (1988)

[6]J.J. Brennan, Interfaces in fiber reinforced glass matrix composites: a case history, in Procs. 2nd European Colloquium on Designing Ceramic Interfaces, Petten, The Netherlands, 11-13 November 1993, p.241-65 (1993)

[7]F. Saugnac, F. Teyssandier, A. Marchand, Characterization of C-B-N solid-solutions deposited from a gaseous-phase between 900°C and 1050°C, *J. Amer. Ceram Soc.*, **75**, p.161-69 (1992)

[8]C. Racault, F. Langlais, R. Naslain, Solid-state synthesis and characterization of the ternary phase Ti_3SiC_2, *J. Mater. Sci.*, **29**, p. 3384-92 (1994)

[9]S. Jacques, H. Fakih, $(SiC/Ti_3SiC_2)_n$ Multilayered coatings deposited by CVD, *Adv. Sci. Technol.*, **45**, p.1085-9090 (2006)

[10]A. Oberlin, Pyrocarbons, *Carbon*, **40**, p.7-24 (2002).

[11]J.-M. Leyssale, J.-P. Da Costa, C. Germain, P. Weisbecker, G. L. Vignoles, An image-guided atomistic reconstruction of pyrolytic carbons , App. Phys. Lett. 95, 231912 (2 p.) (2009)

[12]J. C. Bokros, Deposition, structure and properties of pyrolytic carbon, *Chemistry and Physics of Carbon,* **5**, p.1-118 (1969)

[13]P. Loll, P. Delhaès, A. Pacault, A. Pierre, Diagramme d'existence et propriétés de composites carbone-carbone, *Carbon,* **15**, p. 383-390 (1977)

[14]A. Oberlin, Carbonization and graphitization, *Carbon,* **22**, p.521-41 (1984)

[15]X. Bourrat, F. Langlais, G. Chollon, G. L. Vignoles, Low Temperature Pyrocarbon: a review, *J. Braz. Chem. Soc.*, **117**, p.1090-95 (2006)

[16]J-M. Vallerot, F. Langlais, G. L. Vignoles, X. Bourrat, Growth of Pyrocarbons (in French) *L'Actualité Chimique* n°295-296, p.57-61 (2006)

[17]R.J. Diefendorf, Carbon/carbon composites produced by chemical vapor deposition, in Mechanical Properties and Performance of Engineering Ceramics II: *Ceramic Engineering and Science Proceedings,* **27**, 2, p.399-414 (2007)

[18]R.J. Diefendorf, in J. W. Mitchell, R. C. de Vries, R.W. Roberts and P. Cannon (eds.) Reactivity of solids (Proceedings of the 6th International Symposium on the Reactivity of Solids, Schenectady, New-York, USA August 25-30, 1968), Wiley Interscience, New York, p.461-475 (1969)

[19]R. J. Diefendorf, "Chemical Vapor Deposition of Carbon Matrix Material", in Procs. Carbon Composite Technology Symposium, ASME/Univ. New Mexico, Albuquerque NM, 1970, p.127-142 (1970)

[20]P. Lieberman, H.O. Pierson, Effect of gas phase conditions on resultant matrix pyrocarbon in carbon/carbon composites, *Carbon*, **12**, p.233-41 (1974)

[21]H.O. Pierson and P. Lieberman, "The chemical vapor deposition of carbon on carbon fibers, *Carbon*, **13**, p.159-166 (1975)

[22]R.J. Diefendorf, W.E.Tokarsky, The relationships of structure to properties in graphite fibers Part I. US Air Force report; AFML-TR-72-133, (1972)

[23]X. Bourrat, A. Fillion, R. Naslain, G. Chollon and M. Brendlé, Regenerative laminar pyrocarbon, *Carbon*, **40**, p.2931-45 (2002)

[24]J-M. Vallerot, X. Bourrat, A. Mouchon, G. Chollon, Quantitative structural and textural assessment of laminar pyrocarbons through Raman spectroscopy, electron diffraction and few other techniques, *Carbon*, **44**, 9, p.1833-44 (2006)

[25]B. Reznik, D. Gerthsen, K.J. Hüttinger, Micro- and nanostructure of the carbon matrix of infiltrated carbon fiber felts, *Carbon*, **39**, p.215-29 (2001)

[26]V. de Pauw, B. Reznik, S. Kalhöfer, D. Gerthsen, Z. J. Hu, K. J. Hüttinger, Texture and nanostructures of pyrocarbon layers deposited on plane substrates in a hot-wall reactor. *Carbon*, **41**, p.71-77 (2003)

[27] Y. Sohda, R. J. Diefendorf, Chemical Vapor Infiltration of carbon into graphite blocks with uniform bore capillary tubes – Temperature and pressure as CVI parameters, *Tanso*, **179**, p.153-158 (1997)

[28] Y. Sohda, R. J. Diefendorf, Chemical Vapor Infiltration of carbon into graphite blocks with uniform bore capillary tubes – The effect of feed gas flow and geometrical parameters, *Tanso*, **182**, p.109-114 (1998)

[29]Y. Sohda, R. J. Diefendorf, The Chemical Vapor Deposition of carbon in open-ended capillary tubes, in Procs. 17th biennial conf. on carbon, Univ. Kentucky, Lexington, KY, June (1985)

[30]A. Becker, K. J. Hüttinger, Chemistry and kinetics of chemical vapour deposition of pyrocarbon: V. Influence of reactor volume/deposition surface area ratio. *Carbon*, **36**, p.225-232 (1998)

[31]K. J. Hüttinger, CVD in hot wall reactors—The interaction between homogeneous gas-phase and heterogeneous surface reactions, *Adv. Mater. CVD*, **4**, 4, p.151-158 (1998)

[32]J. Y. Ofori, S. V. Sotirchos, Optimal pressures and temperatures for Isobaric Isothermal CVI, *AIChE J.,* **42** p.2828-2840 (1996)

[33]S. K. Griffiths, R. H. Nilson, Optimum conditions for composites fiber coating by chemical vapor deposition, *J. Electrochem. Soc.,* **145,** p.1263-72 (1998)

[34]P. McAllister, E. E. Wolf, Simulation of a Multiple Substrate Reactor for Chemical Vapor Infiltration of Pyrolytic Carbon Within Carbon-Carbon composites, *AIChE J.,* **39** p.1196-1209 (1993)

[35]T. L. Starr, A. W. Smith, 3D modeling of forced flow thermal gradient CVI for ceramic composite fabrication, *Mat. Res. Soc. Symp. Proc.*, **168**, p.55-60 (1990)

[36]D. Leutard, G. L. Vignoles, F. Lamouroux, B. Bernard, Monitoring Density and Temperature in C/C Composites Processing By CVI with Induction Heating, *J. Mater. Synth. and Proc.* **9** p.259-73 (2002)

[37]G. L. Vignoles, C. Descamps and N. Reuge, Interaction between a reactive preform and the surrounding gas-phase during CVI, *J. Phys. IV France* **10** p.Pr2-9 – Pr2-17 (2000)

[38]N. Reuge, PhD thesis, Université Bordeaux 1, n° 2533 (2002)

[39]G. L. Vignoles, Modelling of CVI Processes, *Adv. Sci. Technol.*, **50**, p. 97-106 (2006)

[40]K.M. Sundaram, G.F. Froment, Kinetics of coke deposition in the thermal cracking of propane, *Chem. Eng. Sci.*, **34**, p.635-644 (1979)

[41]R. Zou, Q. Lou, H. Liu, F. Niu, Investigation of coke deposition during the pyrolysis of hydrocarbon, *Ind. Eng. Chem. Res.*, **26**, p.2528-32 (1987)

[42]L.F. Albright, J.C. Marek, Mechanistic Model for Formation of Coke in Pyrolysis Units Producing Ethylene, *Ind. Eng. Chem. Res.*, **27**, p.755-59 (1988)

[43]D.B. Murphy, R.W. Carroll, Kinetics and mechanism of carbon film deposition by acetylene pyrolysis, *Carbon*, **30**, p.47-54 (1992)

[44]S. Bammidipati, G.D. Stewart, J. R. Elliot Jr., S.A. Gokoglu, M.J. Purdy, Chemical Vapor Deposition of Carbon on Graphite by Methane Pyrolysis, *AIChE J.*, **42**, p.3123-32 (1996)

[45]C.J. Chen, M.H. Back, Simultaneous measurement of the rate of formation of carbon and of hydrocarbon products in the pyrolysis of methane, *Carbon*, **17**, p.175-180 (1979)

[46]J.Y. Lee, J.H. Je, H.S. Kim, A Study on the Properties of Pyrolytic Carbons Deposited from Propane in a Tumbling and Stationary Bed between 900-1230 °C, *Carbon*, **21**, p.523-33 (1983)

[47]P. Lucas, A. Marchand, Pyrolytic Carbon Deposition from Methane: An Analytical Approach to the Chemical Process, *Carbon* **28** p.207-219 (1990)

[48]P. Dupel, R. Pailler, X. Bourrat, R. Naslain, Pulse CVD and infiltration of pyrocarbon in model pores with rectangular cross-sections. 2 – Study of the infiltration, *J. Mater. Sci.*, **29**, p.1056-66 (1994)

[49]W. G. Zhang, K.J. Hüttinger, Chemical vapor deposition of carbon from methane at various pressures, partial pressures and substrate area/reactor volume ratios, *J. Mater. Sci.*, **36**, p.3503-10 (2001)

[50]O. Aubry, PhD Thesis, University of Orleans (2002)

[51]Z. Hu, K.J. Hüttinger, Chemical vapor infiltration of carbon—revised II: Experimental results. *Carbon*, **39**, p.1023-32 (2001)

[52]H.S. Park, W.C. Choi, and K.S. Kim, Process – Microstructure relationships of carbon carbon composites fabricated by isothermal CVI, *J. Adv. Mater.*, **26**, p.34-40 (1995)

[53]F. Fau-Canillac, F. Carrere, A. Reynes, C. Vahlas, F. Maury, Mass Spectrometric Study of the Gas Phase During Chemical Vapor Deposition of Pyrolytic Carbon, *J. Phys IV France*, **5**, C5_89-96 (1995)

[54]A. Becker and K.J. Hüttinger, Chemistry and kinetics of chemical vapor deposition of pyrocarbon, *Carbon* **36**, (1998) pp.177-199, 201-211, 213-224 and 225-232

[55]J. Lavenac, PhD Thesis, Université Bordeaux I, n°2274 (2000)

[56]O. Féron, F. Langlais, R. Naslain, J. Thebault, On kinetic and microtextural transitions in the CVD of pyrocarbon from propane, *Carbon*, **37**, p.1343-1353 (1999)

[57]O. Féron, F. Langlais, R. Naslain, In-situ analysis of gas phase decomposition and kinetic study during carbon deposition from mixtures of carbon tetrachloride and methane, *Carbon*, **37**, p.1355-1361 (1999)

[58]H. Le Poche, X. Bourrat, M.-A. Dourges, G. L. Vignoles, F. Langlais, Influence of the gas phase maturation on the CVD/CVI process and the micro-texture of laminar pyrocarbon from propane, Proc. High-Temperature Ceramic Matrix Composites 5, The American Ceramic Society, Westerville, OH, p.81-86 (2005)

[59]R.J. Diefendorf, The deposition of pyrolytic carbon, *J. Chim. Phys.* (French) **57** p. 815-821 (1960)

[60]M. Tesner, Formation of dispersed carbon by thermal decomposition of hydrocarbons, in Proc. 7th Symposium (Intl.) on Combustion, Butterworths, London, 1959, p.546-553 (1959)

[61]M. Frenklach, On surface growth mechanism of soot particles, in A. R. Burgess et al. (eds.), Proc. 26th Symposium (Intl.) on Combustion, The Combustion Institute, Pittsburgh, p. 2285-2293 (1996)

[62]M. Guellali, R. Oberacker, M. J. Hoffmann, W. G. Zhang and K. J. Hüttinger, Textures of pyrolytic carbon formed in the chemical vapor infiltration of capillaries, *Carbon*, **41**, p.97-104 (2003)

[63]W.G. Zhang, Z.J. Hu, K.J. Hüttinger, CVI of carbon fiber felt: optimization of densification and carbon microstructure. *Carbon*, **40**, 14, p.2529–45 (2002),

[64]J. Lavenac, F. Langlais, O. Féron, R. Naslain, Microstructure of the pyrocarbon matrix in carbon/carbon composites, *Compos. Sci. and Technol.* **61** p.339-345 (2001)

[65]J. Lavenac, F. Langlais, X. Bourrat, R. Naslain, Deposition process of laminar pyrocarbon from propane, *J. Phys. IV France,* **11,** Pr3_1013-1021 (2001)

[66]H. Le Poche, PhD Thesis, University Bordeaux 1, n°2657 (2003)

[67]C. Descamps, G. L. Vignoles, O. Féron, F. Langlais, J. Lavenac, Thermal modelling of a carbon/carbon composite material fabrication process, *J. Phys. IV France,* **11,** p. Pr3_101-Pr3_108 (2001)

[68]C. Descamps, G. L. Vignoles, O. Féron, F. Langlais, J. Lavenac, Correlation between homogeneous propane pyrolysis and pyrocarbon deposition *J. Electrochem. Soc.,* **148,** C695-C708 (2001)

[69]I. Ziegler, R. Fournet, P.M. Marquaire, Pyrolysis of propane for CVI of pyrocarbon: Part I. Experimental and modeling study of the formation of toluene and aliphatic species, *J. Anal. and Appl. Pyrolysis,* **73,** p.212-230 (2005)

[70]I. Ziegler, R. Fournet, P.-M. Marquaire, Pyrolysis of propane for CVI of pyrocarbon: Part II. Experimental and modeling study of polyaromatic species, *J. Anal. and Appl. Pyrolysis,* **73,** p.231-247 (2005)

[71]I. Ziegler-Devin, R. Fournet, P.-M. Marquaire, Pyrolysis of propane for CVI of pyrocarbon: Part III: Experimental and modeling study of the formation of pyrocarbon, *J. Anal. and Appl. Pyrolysis,* **79,** p.268-77 (2007)

[72]N. Birakayala and E. A. Evans, A reduced reaction model for carbon CVD/CVI processes, *Carbon* **40,** p.675-683 (2002)

[73]I. Ziegler, R. Fournet, P.-M. Marquaire, Influence of surface on chemical kinetic of pyrocarbon deposition obtained by propane pyrolysis, *J. Anal. and Appl. Pyrolysis,* **73,** 1, p.107-115 (2005)

[74]R. Lacroix, R. Fournet, I. Ziegler-Devin, P.-M. Marquaire, Kinetic modeling of surface reactions involved in CVI of pyrocarbon obtained by propane pyrolysis, *Carbon,* **48,** 1, p.132-144 (2010)

[75]N. Reuge, G. L. Vignoles, H. Le Poche, F. Langlais, Modelling of pyrocarbon chemical vapor infiltration *Adv. Sci. Technol.,* **36,** p.259-266 (2002)

[76]G. L. Vignoles, C. Gaborieau, S. Delettrez, G. Chollon, F. Langlais, Reinforced carbon foams prepared by chemical vapor infiltration: a process modeling approach, *Surf. Coat. Technol.,* **203,** p.510-515 (2008)

[77]G. L. Vignoles, F. Langlais, N. Reuge, H. Le Poche, C. Descamps, A. Mouchon, From pyrocarbon CVD to pyrocarbon CVI, *ECS Proceedings* **PV 2003-08** p.144-154 (2003)

[78]W. G. Zhang, and K.J. Hüttinger, Inside-Outside Densification of Carbon Fiber Preforms by Isothermal, Isobaric CVI, *ECS Proceedings* **PV 2003-08** p.549 (2003)

[79]G. L. Dong, K. J. Hüttinger, Consideration of reaction mechanisms leading to pyrolytic carbon of different textures, *Carbon,* **40,** p.2515-2528 (2002)

[80]G. L. Vignoles, F. Langlais, C. Descamps, A. Mouchon, H. Le Poche, N. Bertrand, N. Reuge, CVD and CVI of pyrocarbon from various precursors, *Surf. Coat. Technol.,* **188-189,** p.241-249 (2004)

[81]Bourrat X., Lavenac J., Langlais F., Naslain R., The role of pentagons in the growth of laminar pyrocarbon, *Carbon,* **39,** p.2376-80 (2001)

[82]G. L. Vignoles, W. Ros, G. Chollon, F. Langlais, C. Germain, Quantitative validation of a multi-scale model of pyrocarbon chemical vapor infiltration from propane, submitted, *Ceramic Engineering and Science Proceedings* (2012)

HIGH VOLUME PRODUCTION FOR HIGH PERFORMANCE CERAMICS

William J. Walker, Jr.
Federal-Mogul Corporation
Plymouth, MI, USA

ABSTRACT

Ceramic insulators for spark plugs are produced in volumes exceeding 4 million parts per day worldwide, and must reliably provide a high level of performance in a severe service environment. Typically made from alumina, spark plug insulators provide the dielectric barrier to isolate high voltage required for ignition in internal combustion engines. Additionally they act as the mechanical support for the metallic components of the spark plug. With advances in engine technology, the insulators must withstand increasingly harsh conditions and must deliver reliable performance over constantly increasing service lifetimes. Dielectric breakdown of the alumina insulator results in misfiring of the engine, thus dielectric breakdown strength is a critical property of the ceramic and processing must be well-controlled in order to ensure a high degree of reliability. Because of the high production volumes and low cost requirements, dry powder compaction remains the preferred method for manufacturing spark plug insulators. However, processing-related defects may limit the dielectric breakdown strength of these ceramic components. Consequently, processing must be optimized to ensure that performance improvements can be achieved.

INTRODUCTION

The manufacture of spark plug insulators is among the largest uses of alumina in the electronics field. Worldwide, over 4 million spark plug insulators are produced each day which are used in automobiles, small engines and stationary industrial engines. The insulators serve as both a dielectric barrier to high voltage and a structural element. Figure 1 shows typical construction of a spark plug. The insulator is mounted in a steel shell and supports the electrically conducting elements of the spark plug that transmit electrical current into the combustion chamber where a controlled spark between the center electrode and the ground electrode initiates combustion of the air-fuel mixture. The insulators are made from passive dielectric materials, meaning that the ceramic acts as a barrier to the passage of electrical current, even at high applied electrical fields. Dielectric breakdown is the failure – in the form of a physical puncture of the ceramic – caused by an electrical field that exceeds the limit of the material. When a dielectric failure occurs, a secondary path to ground is formed which results in poor engine performance.

Each new generation of automotive engine technology results in increased operating demands on spark plugs. In general, today's engines operate at higher temperatures and require higher voltages in order to deliver increased performance and economy. These factors, in combination with the demand for more compact plug designs and ever-longer service life, continually push the capabilities of the materials. At the same time, the available "footprint" for the spark plug is shrinking due to the increased sizes of cooling passages and valves. Just a few years ago, automotive spark plugs typically used a 14 mm thread to mount in the engine. Today, 12 mm is becoming the norm and 10 mm is expected in the next few years, with the diameter of the insulator reducing by corresponding amounts (see Figure 2). Moreover, the utilization of leaner fuel mixtures, increased compression and/or alternative fuels depends on available energy to ignite the mixture. In spite of these rising demands, vehicle manufacturers and other customers expect longer plug life, with the ultimate target being the lifetime of the engine.

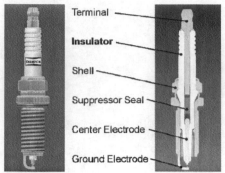

Figure 1. Overview of spark plug construction.

Spark plug manufacturers have addressed these demands through design changes and the use of longer-lasting precious-metal electrodes, such as iridium. One of the remaining barriers to increased plug performance and durability, however, is the ceramic used in the plug insulator. Being the heart of the spark plug, the ceramic insulator isolates the high voltage applied to the terminal of the spark plug from the grounded shell. With increased applied voltages, higher operating temperatures and thinner insulator cross-sections, existing ceramic materials are reaching their practical limits. Improvements in ceramic technology are critical to developing higher-performance ignition systems. As engine technology progresses, the role of the ceramic spark plug insulator in controlling voltage becomes increasingly important.

Consumers clearly are demanding increased power and overall performance and durability, as well as increased fuel efficiency from automotive engines. These needs have helped drive the downsizing of under-hood compartments and engine packages, thereby affecting spark plug geometries. These new size and shape restrictions redouble the need for improved ceramic materials.

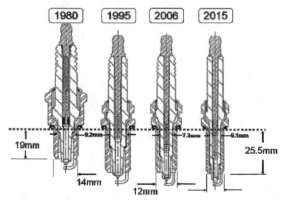

Figure 2. Evolution of spark plug design. Increases in the size of valves and cooling passages in engines limits the available space for spark plugs, necessitating smaller diameter insulators with reduced thickness.

MANUFACTURING SPARK PLUG INSULATORS

Spark plug insulators are almost universally formed by a dry-bag isostatic pressing process. The manufacturing sequence is shown schematically in Figures 3 and 4. The starting materials, alumina powder and minor amounts of kaolin, talc and limestone, are finely ground in an aqueous suspension or slurry. A small amount of a binder-lubricant is added and the slurry is spray-dried by atomization in a heated chamber, forming a free-flowing, granulated feed material for pressing. The spray-dried powder is fed into rubber molds and formed into insulator blanks by using isostatic pressure to compact the powder around steel arbors which form the bores of the insulators. The outer surfaces of the insulators are formed by profile grinding in the green state using shaped abrasive wheels. The parts are then fired to about 1600°C to sinter the powder into a dense ceramic. Subsequent processing adds decoration and glaze, which are fired to a lower temperature.

PROPERTY REQUIREMENTS FOR SPARK PLUG INSULATORS

The insulator for a spark plug is generally cylindrical with an axial bore, and is mounted in a metal shell which threads into the engine block. The high-voltage lead is connected to a metal terminal projecting from the end of the bore away from the combustion chamber. An electrode projects from the other end of the bore into the combustion chamber. This electrode is typically a nickel alloy with a copper core, and may be tipped with platinum, iridium or some other precious metal alloy. The electrode and terminal are connected by means of an electrically resistive glass seal, which has the combined functions of forming a gas-tight seal to prevent passage of combustion gasses and providing sufficient electrical resistance to suppress radio frequency noise from the spark. The outer profile of the insulator is contoured to have an enlarged section to lock the insulator in the shell, and a tapered tip or "nose" that extends beyond the end of the shell into the engine. The design of the nose controls the heat range of the spark plug and prevents tracking of the spark along the surface of the insulator to the shell. The portion of the insulator that extends out of the engine to the terminal is generally cylindrical and is coated with a glaze that prevents buildup of dirt that can cause flashover to occur.

In order to withstand the severe environment of a modern engine, the ceramic must have the following properties:

(1) High Dielectric Strength – Ignition voltages can exceed 40,000 volts.
(2) High Electrical Resistance – Leakage currents reduce the power to the spark, and degrade the long-term performance of the insulator.
(3) High Mechanical Strength – The insulator is mechanically locked in a metal shell, and is subjected to vibration and cyclical high pressure loading of as much as 20 MPa.
(4) Resistance to High Temperatures – The tip of the insulator can reach 1000°C.
(5) High Thermal Conductivity – The spark plug insulator contributes to the conduction of heat from away from the firing end in order to avoid becoming a source of pre-ignition, while retaining enough heat to burn off conductive deposits.
(6) Good Thermal Shock Resistance – The ceramic must withstand the rapid change from a cold start to the operating temperature of an engine, and withstand the constant thermal cycling between impingement of the cool fuel-air mixture and combustion temperatures.
(7) Chemical Inertness – The ceramic cannot degrade due to exposure to corrosive combustion gasses and fuel additives.

Dielectric breakdown strength is of particular importance in today's engines. Figure 5 shows the dielectric loading on an insulator. The maximum field strength occurs where the insulator seats in the shell. A secondary maximum occurs at the head of the center electrode.

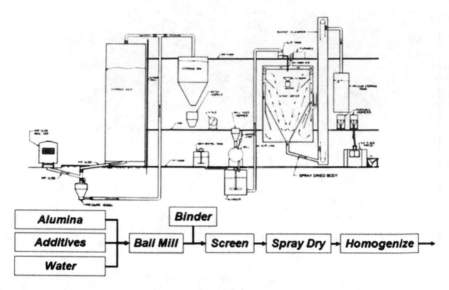

Figure 3. Materials preparation for spark plug insulator manufacture.

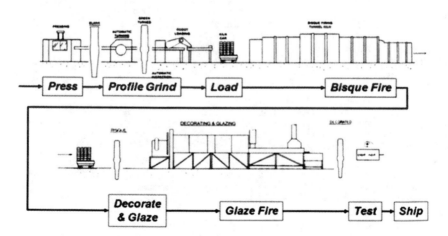

Figure 4. Manufacturing process for spark plug insulators.

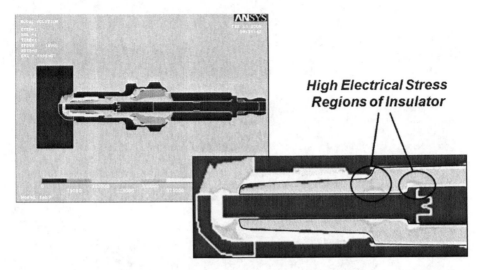

High Electrical Stress Regions of Insulator

Figure 5. Electric field modeling shows that the maximum electric field occurs at transitions in the internal and exterior profile of the insulator in close proximity to contact points with the shell and center electrode.

In addition, cost is a key consideration for the materials used for spark plug insulators. Very few ceramic materials combine the necessary properties for spark plug insulators, and of those, only one is cost effective. The ceramic used for spark plug insulators is comprised primarily of alumina. Alumina is produced in enormous quantities from bauxite using the Bayer process as an intermediate step in the production of aluminum metal. A small portion of that material is used for alumina chemicals and ceramics.

Alumina is a very hard, strong, chemically inert material that has been the basis for spark plug insulators since the advent of leaded gasoline in the 1930s[1]. Alumina is blended with small amounts of kaolin, talc and limestone, which react at high temperature to create a liquid phase that aids in consolidating the material during sintering and persists as a glass in the final ceramic material. Early alumina ceramics used for spark plug insulators contained about 85% alumina and 15% glass. The glass lowers the firing temperature of the ceramic, resulting in less costly processing, but also lowers the electrical and mechanical properties of the material. The trend over time has been to use more alumina and less glass. Today's spark plugs use ceramic compositions of 95 to 96% alumina. Figure 6 shows typical microstructure of alumina ceramic used for spark plug insulators.

DIELECTRIC BREAKDOWN
Dielectric breakdown occurs when an insulation material fails to insulate due to a rapid increase in charge carriers passing through the material once the applied electric field exceeds some critical level. In the case of gasses, this occurs when an arc forms between electrodes. A spark plug is designed to apply electric fields that exceed the dielectric strength of the gaseous mixture of air and fuel in the cylinder of an engine in a reliable and consistent manner in order to control ignition. The insulator, however, must be capable of withstanding these same electric fields.

Dielectric breakdown of the insulator results in a physical puncture through the material (see Figure 7). Breakdown may occur due to one of three mechanisms[2]. Intrinsic or electronic breakdown occurs at very high electric fields where electrons may be ejected from the electrode into the insulating material, or electrons may be promoted across the band gap from the valence band to the conduction band. Some of these electrons collide with atoms or electrons, generating more electrons, resulting in an avalanche of conducting electrons concentrated in a small area. Intrinsic breakdown is rarely observed in solids because of the high electric fields that are required. Thermal breakdown is the most common mechanism of dielectric breakdown in ceramic insulators, occurring at electric fields about two orders of magnitude lower than those required for intrinsic breakdown. Movement of charge carriers within the material in response to the applied electric field results in localized heating, which in turn results in additional movement of charge carriers and more localized heating. The resulting thermal runaway reduces the voltage required to initiate the avalanche of conducting electrons that cause dielectric breakdown. Ionization breakdown occurs as a result of the ionization of gasses within pores and cracks of insulators with a non-homogeneous structure. The high-energy ionized gas generates heat locally, which promotes thermal breakdown. Thus the factors affecting dielectric breakdown of a material include the number and mobility of charge carriers and microstructural inhomogeneities such as cracks and pores.

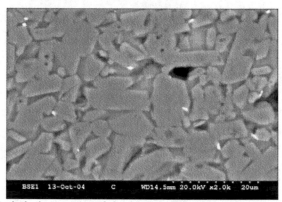

Figure 6. Typical microstructure of alumina ceramic used to produce spark plug insulators.

Dielectric breakdown strength is statistical in nature and can be described using Weibull statistics[3]. Consider a volume of ceramic that is subjected to a uniform electric field. Dielectric breakdown will initiate due to the thermal and/or ionization breakdown mechanisms in a region where the electric field is concentrated due to microstructural features. Garboczi[4] has proposed a Griffith-like criterion for the initiation of dielectric breakdown from electrical field concentration due to microstructural defects. In this model the critical electric field strength E_e^0 necessary to initiate dielectric breakdown is dependent on the size of an elliptical defect with length $2c$ according to the equation

$$E_e^0 = \frac{K_{Ic}}{\sqrt{\pi c}}$$

where K_{Ic} is the critical electric field intensity factor.

Figure 7. SEM images of a dielectric puncture through an insulator showing the effects of thermal breakdown. The puncture propagated radially from the center electrode to the shell and generated sufficient heat to melt the ceramic. The insulator was carefully broken to reveal the path of the puncture.

Within the material, there will be a distribution of defects with different sizes. Just as a distribution of strength-limiting defects affects the mechanical performance of ceramic materials, defects affect the dielectric reliability of a population of insulators. The Weibull distribution is described by the equation

$$P_f = 1 - exp\left\{\int_V \left(\frac{E}{E_0}\right)^m dV\right\}$$

where P_f is the probability of failure of a specimen, V is the specimen volume, E is the applied electrical field and E_0 is the characteristic electric field where P_f is 63.2% and m is the Weibull modulus. By fitting dielectric puncture data to the Weibull model, dielectric reliability can be determined.

MEASUREMENT OF DIELECTRIC BREAKDOWN STRENGTH

Dielectric breakdown strength of solids can be measured according to ASTM D149[5] and IEC 60243-1[6] by placing a planar test specimen between two electrodes and increasing the applied electric field until failure occurs. Failure is detected by measuring an increase in the current between the electrodes, and verified by visually locating a puncture. It is usually necessary to immerse the test fixture in a dielectric fluid to prevent flashover between the electrodes. A modification of this test uses a tubular test specimen[7] which is formed using the same equipment as is used to make spark plug insulators. Specimens are tested with a high-voltage lead connected to a rod electrode positioned on the inside and ground ring on the outside of the specimen. A 60 hz AC electric field is increased at a rate of 500 V(rms)/sec until failure occurs. Automotive companies generally prefer dielectric testing performed on assembled spark plugs using an automotive ignition source in order to test the performance of the device rather than the property of the material alone. The spark plug is threaded into a pressure vessel which is pressurized with air or some other gas in order to suppress the spark between the electrodes and force puncture through the ceramic[8]. The high-voltage lead from an automotive ignition coil is attached to the terminal of the spark plug and the pressure vessel is grounded. The voltage signal is a series of DC pulses at 50 hz, and is increased in 1 kV steps in 1 sec. intervals until puncture is detected by a change in the waveform of the voltage and verified by inspection. Sets of 30 or more spark plugs ensure reliable statistics.

Figure 8 shows Weibull plots of the dielectric breakdown voltage of two sets of spark plugs processed under different conditions. Set A was produced under less than optimal conditions. At the low end of the distribution, two points deviate from the best-fit line for the Weibull distribution. The lowest point is outside the 95% confidence interval range for the Weibull distribution, suggesting that this low flier was caused by a defect that is due to a different type of defect than those in the rest of the population. The poor performance of this one specimen severely limits the dielectric reliability of spark plugs made with those insulators. In contrast, Set B was produced under improved conditions. The Weibull modulus was increased from 26.5 to 31.0 and the characteristic puncture voltage was increased from 41.8 kV to 43.3 kV, but more importantly, the low fliers were eliminated. The reliability of the insulators is significantly improved.

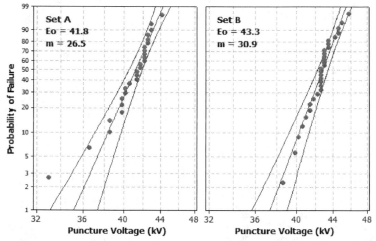

Figure 8. Weibull plots with 95% confidence interval showing dielectric performance of two sets of spark plugs using insulators produced under different conditions. Testing was conducted on complete spark plugs using an automotive ignition coil to provide the voltage, using pressurized gas to suppress normal sparking at the firing tip in order to force puncture through the ceramic insulator.

SUMMARY

Advances in automotive engine technology push the demands on spark plugs, necessitating more compact plug designs that operate under more severe thermal and electrical stresses over an ever-longer service life. Ceramic insulators with improved properties are needed to meet these demands, and need to be manufactured on a scale of millions of parts per day in order to supply the automobile industry. Insulators for spark plugs are almost universally made from aluminum oxide ceramic and formed by a dry bag compaction process. The ceramic insulator provides the dielectric barrier to isolate the high voltage required for ignition, and acts as the mechanical support around which the spark plug is built. Dielectric failure of an insulator results in poor engine performance so the insulators must be manufactured with a high degree of reliability. The microstructural defects that cause low dielectric breakdown strength in a small number of insulators can be eliminated by carefully controlling the manufacturing process. Weibull statistics are a useful tool for ensuring reliability of dielectric performance.

REFERENCES
[1] J.S. Owens, J.W. Hinton, R.H. Insley and M.E. Poland, "Development of Ceramic Insulators for Spark Plugs," American Ceramic Society Bulletin, 56 [4] 437-440 (1977).
[2] R.C. Buchanan, "Ceramic Insulators," pp. 1-86 in Ceramic Materials for Electronics, 3rd Edition, R.C. Buchanan, editor. Marcel Dekker, Inc., New York, 2004.
[3] Y. Wang, Y.C. Chan, Z.L. Gui, D.P, Webb and L.T. Li, "Application of Weibull Distribution Analysis to the Dielectric Failure of Multilayer Ceramic Capacitors," Materials Science and Engineering B47, 197-302 (1997).
[4] E.J. Garboczi, "Linear Dielectric-Breakdown Electrostatics," Physical Review B, 238 9005-9010 (1988).
[5] ASTM D149-97a (2004) Standard Test Method for Dielectric Breakdown Voltage and Dielectric Strength of Solid Electrical Insulating Materials at Commercial Power Frequencies.
[6] IEC 60243-1: Electrical Strength of Insulating Materials - Test Methods - Part 1: Tests at Power Frequencies.
[7] R.H. Insley, "Electrical Properties of Alumina Ceramics," pp. 293 – 298 in Alumina Chemicals, L.D. Hart, Editor, The American Ceramic Society, Inc., Westerville, Ohio, 1990.
[8] JIS B8031-2006 Japanese Industrial Standard for Internal Combustion Engines-Spark Plugs.

LOW PRESSURE INJECTION MOLDING OF ADVANCED CERAMIC COMPONENTS WITH
COMPLEX SHAPES FOR MASS PRODUCTION

Eugene Medvedovski
Umicore TFP
Providence, RI, USA

Michael Peltsman
Peltsman Corp.
Minneapolis, MN, USA

ABSTRACT
This paper reviews the technology of low pressure injection molding that is successfully used for manufacturing of advanced ceramics with different compositions, particularly for the components with custom-designed complex shapes when the production is achieved hundreds pieces/ day or week or higher. The major principles of this technology are reviewed based on the extensive processing experience. Semi-automated equipment for low pressure injection molding technology is described.

INTRODUCTION
Among the variety of advanced ceramic components for structural, high-temperature and corrosion resistant applications, electrical insulators, semiconductors and others, many of them have complex, difficult for production, shapes. Manufacturing of these complex shape components is especially challenging when their output is quite large, e.g. hundreds of pieces per week or per day. In this case, widely used manufacturing methods, such as slip casting, uniaxial or cold isostatic pressing, are not very suitable because the mechanical treatment (grinding) of green or fired bodies becomes very expensive, not productive, and it creates additional mechanical stresses, which significantly reduce the reliability of the ceramics. The alternative high-productivity methods, which allow to manufacture complex shape bodies with accurate dimensions with minimal processing waste, have to be utilized. Injection molding technology is one of the good possibilities to accomplish these tasks[1-7].

Injection molding technology is based on the ability of ceramic mixtures, which have the consistency of slurries prepared with a specific temporary polymer binder and plasticizer system and heated to the certain temperature, to flow under certain pressures filling the cavity of metallic molds. When the slurry is cooled down in the mold, it is solidified resulting in formation of a green body. There are two principle methods of injection molding technology actively used in ceramic industry. They are distinguished mostly by the type of the temporary binder system and its properties and by the related pressure applied. Due to these differences, equipment used for the shaping of ceramic components and method of the binder removal are also distinguished.

The first method named as high pressure injection molding is based on the use of the thermoreactive organic compounds, which become fluid at temperatures of 150-300°C. In this case, a ceramic powder is plasticized with this binder system at the mentioned temperature range with consequent granulation, and then the heated plasticized ceramic-polymer compound is fed into the injection machine with a piston; the shaping is conducted under rather high pressures (5-70 MPa) into metallic molds[1-7]. The obtained green body is heat treated for the binder burnout with consequent firing.

Another method named as low pressure injection molding is based on the use of thermoplastic organic compounds, which become fluid at low temperatures, such as 60-70°C. The major component of this binder system is paraffin-wax that melts at this low temperature. Because the ceramic-polymer compositions based on paraffin have rather low viscosity and good fluidity, very high softness and plastic properties at rather low temperatures, these compositions require only low pressures (0.2-0.7

MPa) to force them to fill the mold cavities. In this case, a ceramic powder is mixed and plasticized with this paraffin-based binder system at 60-70°C and the prepared composition is injected (cast) into metallic molds. When the mould is cooled down, a solidified green body is ejected from the mold. After the binder removal (debinding or dewaxing), a ceramic body is fired for the required density. This method and its major principles were developed in the 1950-1960-s by P.O. Gribovsky[8]; at that time, this method was named as "hot casting" or "hot casting of thermoplastic slurries". Optimization of this technology allowed to produce different ceramic components for various applications on the high-volume industrial basis for years[8-14]. This technology is reviewed in the present paper based on the vast studies and practical manufacturing experience for different ceramic materials, including in the mass-production environment. Manufacturing equipment designed and fabricated is described. Some important processing features, which affect quality of ceramics and processing yield, are pointed out.

MAJOR PRINCIPLE OF LOW PRESSURE INJECTION MOLDING TECHNOLOGY

Low pressure injection molding technology is applicable for a variety of ceramic materials, including oxide and non oxide ceramics contained one, two or several different phases and ingredients. This technology allows to produce components for structural and electrical applications, e.g. based on alumina, zirconia, spinel, mullite, titania, silicon nitride, silicon carbide, steatite, cordierite, forsterite, spodumene, celsian, different electroceramics (semiconductors, piezoceramics and segnetoceramics, etc., e.g. titanates, zirconates, stannates) and many others, as well as cermets. These materials and products are widely used for wire and thread guides in cable and textile industries, respectively, various wear- and corrosion-resistant components for oil & gas and mineral processing (e.g. valves and seats, pump components, impellers, etc.), cutting tools, bearings, high-temperature nozzles and thermal shock resistant supports, electrical insulators for high-frequency applications, electro-vacuum devices and feedthroughs for nuclear powder plants and reactors, spark plug insulators, heat engine components, biomedical components (e.g. implants, joints, orthodontic brackets) and many others.

The manufacturing process includes a number of steps, which are schematically presented in Fig. 1. As the first step, the process includes preparation of the solid ceramic ingredient in the powder form. In some cases, if the ceramic ingredient is based on two or more components, which may have phase transformation during consequent thermal processes and if special dopants have to be used for sinterability improvement or physical properties modification, the ceramic ingredient has to be preliminary synthesized to eliminate the stresses dealt with the phase transformation and related volume changes or to reduce fired shrinkage. In any case, the mineral components of the solid ceramic ingredient have to be very uniformly distributed, regardless if the preliminary synthesis is required or not (for instance, when two ceramic oxides without phase transformation or oxide – non-oxide or cermets are manufactured). The ceramic ingredient, including the synthesized material, has to be disintegrated to obtain a powder with certain parameters, which are necessary for appropriate manufacturing and sintering.

The prepared powder is mixed with the temporary paraffin-based binder system at elevated temperature (65-80°C, sometimes up to 100°C) when this binder system becomes liquid, and this mixing should provide uniform distribution of the solid-liquid ingredients. In order to eliminate the presence of air bubbles in the ceramic-polymer slurry, vacuuming during the mixing step is necessary. Due to a reversible liquid-solid transformation of the paraffin-based binder and related slurries, the ceramic powder-binder compositions can be stored in a solid form (i.e. feedstock), and they can be melted and used for the consequent processing steps. The slurry with thermoplastic properties is injected under pressure applying compressed air into metallic custom-designed molds providing required shapes, and then the formed bodies are removed from the molds when they become cold. The principle schematic of the injection molding device is shown in Fig. 2. The cycle of injection-cooling-demolding-mold assembling is rather fast (from less than 1 min. to several min.) that allows to produce up to several hundred pieces per working shift. Ceramic bodies obtained are mechanically very strong, and they can be handled without breakage, even the pieces with very thin walls with holes or with uneven wall thickness. If it is necessary, they can be machined on a lathe or drilled. The obtained near-net shape components are heat treated to remove the binding components, and then final firing is conducted up to fully sintering conditions or to the state of required porosity. Due to the ability of the paraffin-based materials to become liquid at low temperatures without degradation, the waste of the thermoplastic slurry or defected cast components in a green state can be returned to manufacturing, i.e. this technology is low-waste that is especially beneficial for processing of expensive materials.

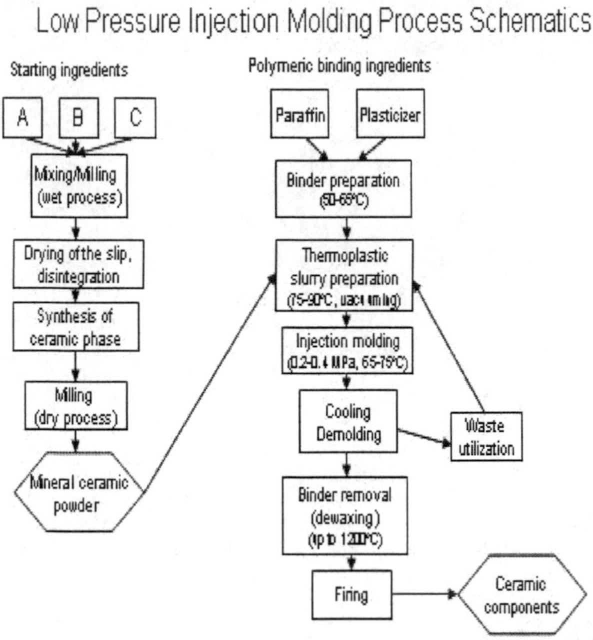

Low Pressure Injection Molding Process Schematics

Fig. 1. Schematics of the Low Pressure Injection Molding Processing

Different factors, including morphology and properties of ceramic powder (solid phase) and its preparation technologies, composition of thermoplastic binder system, ratio of solid (ceramic powder)/liquid phase (binder system), thermoplastic slurry parameters, injection process, mold design, debinding and final firing process and other features, affect quality and processing yield of low pressure injection molded ceramics. Some key processing factors affecting quality of ceramics are considered below. This consideration is important in order to maintain a high processing yield in manufacturing and high quality of the produced ceramics, especially in mass-production environment.

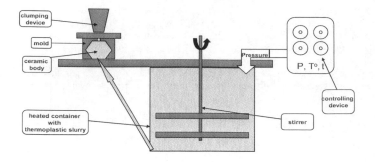

Principle Schematic of Low Pressure Injection Molding Device

Fig. 2. Principle Schematics of the Low Pressure Injection Molding Device

SOME FACTORS AFFECTING LOW PRESSURE INJECTION MOLDING PROCESSING
Mineral Ceramic Powder (Solid Phase) Preparation Features

Features of ceramic powders used for low pressure injection molding, such as phase composition, morphology, particle size and some others, as well as the method of their preparation, significantly affect the wetting of the solid phase by the liquid binder and the required content of the liquid phase in the thermoplastic slurry, and, as a result, slurry properties, green body compaction and processing yield[8-11, 15]. A number of studies conducted on the laboratory basis consider the use of "single-compound" powders, e.g. pure alumina, zirconia and some other oxides without mineral additives. However, in many industrial applications, the ceramic compositions have to contain special additives - dopants promoted ceramic densification and required properties (mechanical, electrical, thermal and others). Many electroceramics, electroinsulating materials and structural ceramics, e.g. perovskites, spinels, mullite-based, steatite, cordierite, many industrial types of alumina ceramics and others, are manufactured using several starting mineral ingredients, which (or some of them) have phase transformation with related sufficient volume changes and gas removal. In all these cases, synthesis of the ceramic powders is a necessary step to achieve a high homogeneity of ceramics, and its conditions have a significant influence on the low pressure injection molding process.

Phase composition, structure and morphology of these ceramic powders are defined, in a high extent, by the temperature of synthesis of the ceramic phase. Based on the literature data[8, 9] and based on practical experience, when temperature of synthesis is higher (i.e. when the synthesis is completed), the specific gravity of the powder is higher with certain morphological features and, generally, lower contents of the liquid phase may be required to obtain the slurries with workable parameters. Only 8-15 wt.-% of the binding-plasticizing ingredient (depending on the type of ceramics) is required for the slurry preparation, and this lower content of the liquid phase positively affects the debinding process and shrinkage reduction. The examples of the influence of the temperature of synthesis for some ceramic powders on the content of required binder system are shown on Fig. 3. In comparison with these data, Lin and German[16] used more than 40 wt.-% of the binder system for the preparation of the thermoplastic slurry based on the alumina powder doped with MgO synthesized at only 800°C; this high amount of the organic ingredient will definitely result in processing difficulties in debinding, elevated shrinkage, etc. However, if temperature of synthesis of ceramics is too high, the synthesized powder becomes very hard, and longer milling is required that may result in the rise of binder needed and occurred difficulties with debinding.

Optimization of the milling technology of the synthesized ceramic phase allows to obtain the powders with high specific surface area and low particle size (micron or sub-micron powders), i.e. with adequate sinterability, and which require minimal binder system contents. Highly stable slurries are usually obtained from the powders with a high level of homogeneity and a certain level of specific surface area (e.g. from 0.3-0.5 to 3-5 m²/g or higher depending on the ceramics) [8-10, 17]. It is very important to obtain the powders with absence of hydrate layers on their surface, i.e. with high wetting of these powders by non-polar thermoplastic binders. As indicated in old literature sources[8, 9] and observed and established in industrial condition (including by the authors of the present paper), that the presence of even small amounts of hydroxyl groups on the ceramic surface inhibits the creation of hydrogen bonds between hydrocarbons of the thermoplastic binder systems and oxygen atoms of the crystalline lattice of the solid ceramic phase; this results in the increase of the binder content. Similar results were also described later by S. Novak et al[15]. In order to exclude the noted problems, right milling equipment and technology, as well as right storage of the powders and feedstocks, have to be used. In order to facilitate a milling process of the ceramic phase, small amounts of some surfactants may be successfully used.

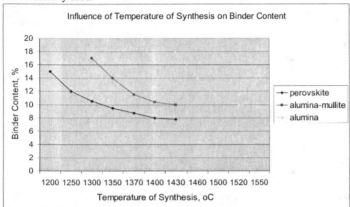

Fig. 3. Influence of Temperature of Synthesis of Ceramic Powders on the Binder Content of Thermoplastic Alumina, Alumina-Mullite and Perovskite Ceramic Slurries

Binder System for Low Pressure Injection Molding

In order to obtain ceramic slurries with minimal contents of a liquid phase but with low viscosity, to attain good filling of the mold cavity and good particles packing with a possible reduced shrinkage, a liquid phase (e.g. a binder system) has to have a good adhesion to ceramic particles. Although the excessive binder content lowers viscosity of the slurries and provides better mold cavity filling, it may cause the powder-binder separation under the relatively high stress molding, distortion at the demolding, debinding defects and higher fired shrinkage. The binder system should provide high mechanical strength of green bodies, which often have complex shapes with uneven thin-walled sections. The organic binder system used for injection molding usually consists of several components, which, as indicated by M. Edirisinghe[18], includes:

a) a major binder that determines the general range of final binder properties;
b) a minor binder that is a thermoplastic polymer or oil;
c) a plasticizer that is added to increase the fluidity of the ceramic-polymer slurry;
d) processing additives, which are used as surfactants to improve wetting of the ceramic ingredient by the liquid phase.

In the case of the low pressure injection molding slurries, the main binder component is paraffin wax that is a thermoplastic polymer which is a mix of hydrocarbons C_nH_{2n+2} (n=19-35) providing melting and low viscosity at relatively low temperatures. Considering the type of paraffin, it is desirable to use low-molecular weight paraffin waxes [3, 5, 8-15, 19-22]. The paraffin contents in the binder systems may achieve more than 90%. As very common additives, small amounts of other waxes, e.g. bee wax, polyethylene wax, carnauba wax, may be used as plasticizing agents and surfactants, which also have a good affinity with paraffin and provide viscosity reduction.

Steric stabilization of the non-polar suspensions based on paraffin wax can be achieved by the use of special organic additives - surfactants. These additives improve dispersion of ceramic ingredient in the binder system, miscibility between the binder system ingredients and lubrication of the mold[18, 23]. The surfactants consisting of polar radicals and non-polar hydrocarbon chains are adsorbed on the mineral powder surfaces by the polar radical, but the non-polar chains create a "protection" layer for solid particles against moisture. In this case, the stabilized "fatty-like" powders can be easy wetted by non-polar molecules of paraffin hydrocarbons. The content of surfactants-stabilizers needed is very small (even up to 0.1-0.2 wt.-% based on the weight of a ceramic powder). As the surfactants, short-chain organic substances (C_{12}-C_{22}) such as fatty acids, e.g. oleic or stearic acids and some others, and/or some esters may be used, as many authors indicated[8-11, 15, 18-20, 23-26]. Also the additives as some other carboxylic acid (e.g. octadecanoic acid, 12-hydroxystearic acid), octadecylamine, fish oil, silicone were tested[15, 18, 20, 23]. The surfactant additives provide a formation of the -Me-O-CO- bonds, and this esterification significantly reduces the powder agglomeration through steric repulsion and provides the stabilization of the slurry[18-20, 24].

An addition of long-chain carboxylic acids may positively affect the steric stabilization of non-polar suspensions. Particularly, oleic and stearic acids are commonly used for viscosity decrease of the paraffin-based slurries and as surfactants due to reduction of the ceramic powder agglomeration. Many authors[15, 16, 18-20, 23, 26] prefer to use stearic acid as the surfactant, but most of their works were related to high pressure injection molding. However, an addition of stearic acid results in higher residual carbon contents attributed to a strong adsorption onto the ceramic powder surface[21]. Besides, Lin et al[16] and Chan et al[19] noted the possibility of forming bubbles in the slurry arising from evaporation of stearic acid. At the selection of the surfactant, its melting point has a great importance. For example, while oleic acid is a liquid at temperature of 15°C, stearic acid and 12-hyrdostearic acid melt at ~68°C and 85°C, respectively, and the use of "high-temperature" surfactants may create some difficulties in the industrial processing for low pressure injection molding where the "low-temperature" binder system is used. Based on our experience, oleic acid as a surfactant, dispersant and milling "promoter" is very effective in industrial manufacturing.

Modification of the oxide ceramic suspensions is based on steric stabilization by a combination of short- and long-chain molecules adding to non-polar liquids with addition of dispersant and plasticizing agent to the non-polar binder. For example, an addition of short- and long-chain molecules (e.g. fatty amine and saponified wax) provides easier homogenization and increase of solid loading, and an addition of non-polar wax (e.g. polyethylene-wax or bee-wax) increases stability of molded bodies and promotes immobilization of paraffin and solid particles. The adsorption layers for effective steric stabilization should offer not only evolution of powder surface and adhesion, but also provide certain deformation properties under mechanical force applied. R. Lenk et al[22] recommended a combination of short- and long-chain surfactants and plasticizers (fatty amine and alkylsuccinimide, respectively) for the surface modification of SiC. The stabilizing effect of these surfactants resulted in improved flow behavior of the highly-concentrated hot paraffin-based SiC slurries. The dispersion mechanism for the ceramics used for high- and low-pressure injection molding is considered in details in the above mentioned references.

Ceramic paraffin-based slurries used for low pressure injection molding have relatively high solid contents (85-90 wt.-%). At this level of solids, viscosity of the thermoplastic slurries is low enough for good flow and filling the molds. The rheological behaviour of the thermoplastic slurries requires special consideration that is not provided in this paper. Briefly, these slurries have nearly Newtonian behaviour[15, 20] if they have optimal quantity of the suitable surfactant. A small surplus increases the shear stress and pseudoplasticity of the slurries[15, 19, 20, 27]. In general, the relation of viscosity vs. solid content is in accordance with the Krieger-Dougherty model[20, 28, 29]:

$$\eta = \eta_{lp} \left(1 - \Phi/\Phi_m\right)^{-n},$$

where η_{lp} is viscosity of the liquid phase, Φ is a solid volume fraction, Φ_m is maximal packing of solid fraction in the slurry, respectively, and n is an empirical coefficient usually taken as 2-2.5 for spherical particles. Lower viscosity with a noted high solid content may be achieved by the optimized addition and mixing of the binder with solid materials and appropriate handling of the slurry (e.g. by constant stirring), which provide its high homogeneity, by addition of surfactants and by selection and maintaining of proper temperature of the slurry. The considered ceramic slurries have a thixotropic behavior, and this thixotropic behavior increases with increase of a solid content.

Slurry Injection Features and Parameters

Thermoplastic slurries used for low pressure injection molding have to be of a high homogeneity that promotes greatly the yield and consistent properties in the mass-production conditions[8, 9, 17]. It includes a high uniformity of the binder distribution, stability of the slurry, absence of the air bubbles, and it is defined by the maintaining certain technological parameters and utilized processing equipment. Based on practical industrial experience, homogeneity of the slurries, e.g. stable behavior without settling of the solid phase, can be increased not only by mechanical action, e.g. stirring, and by optimization of the ratio of solid/liquid phase, but also by the use of powders with finer particles (with micron and sub-micron particle sizes) obtained at the temperatures when the synthesis of a solid phase is completed. In order to reduce (eliminate) the presence of air bubbles in the thermoplastic slurries and to stabilize their properties, the slurries have to be vacuumized in advance and stored with stirring before the shaping process.

Injection molding parameters significantly affect properties of ceramics and processing yield. The major parameters include injection pressure, speed of injection, time of the holding of pressure during injection, temperature of ceramic slurry, temperature of mold, cooling of mold (area and direction of cooling) and some others. Some parameters have to be selected based on the shape and size of the ceramic body and mold design. For example, even design and size of a feeding sprue affect hardening of the injected ceramic body and, finally, the processing yield. As mentioned above, the properties and processing yield, e.g. absence or presence of defects, are defined by not only the injection molding parameters but these parameters are specially selected and optimized for different ceramic materials. Some examples of the influence of injection molding parameters on green and fired density are shown for high alumina, alumina-mullite and steatite ceramics, which were selected for demonstration among many other materials produced in accordance with the described technology.

An increase of injection pressure promotes particles compaction and green density; higher amounts of the slurry are used to fill the mold cavity. Shrinkage becomes lower accordingly. As a result, firing density is increased. However, an injection pressure increase is effective only up to 0.5-0.7 MPa; then, at higher pressures, density increases insufficiently (Fig. 4).

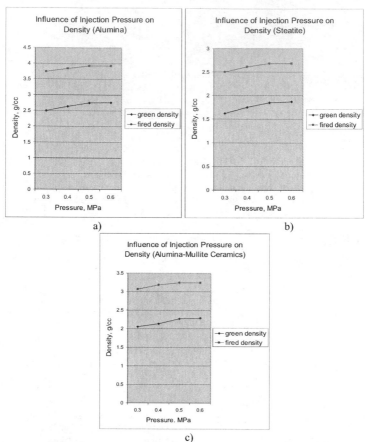

a) b)

c)

Fig. 4. Influence of Injection Pressure on Density of Low Pressure Injection Molded Ceramics (a – steatite, firing temperature 1280°C; b – high alumina ceramics, firing temperature 1650°C; alumina-mullite ceramics, firing temperature 1410°C) Process conditions: temperature of slurry 90°C, temperature of mould 15°C

The influence of speed of injection on the quality of ceramic bodies is not very common. When the injection speed is in the range of 25-100 cm³/sec, compaction is increased, but, at higher speeds, air bubbles may occur in the green bodies due to the turbulence at the injection, even when the slurry was vacuumized, that reduces density of ceramics. At the "extreme" injection speeds, the slurry may reach the top of a mold very fast with formation of air pockets. In this case, the trapped air may even break the ceramic body after release of pressure. However, the injection speed has to be adjusted based on the ceramic body configuration and related mold design. For example, if the molds have the parts preventing the flow of the slurry, i.e. speed of the flow decreases during filling the mold, the injection speed may be elevated, and the possibility of air bubbles trapping is reduced. The "safe" speed of injection is also adjusted based on the slurry flowability that is defined in a significant extent by the slurry parameters, e.g. temperature. The features of the influence of speed of injection and temperature of the slurries are shown on Fig. 5. Analyzing the process, it can be seen, that turbulence of the slurry occurs at lower speeds with the slurry temperature increase due to sufficient reduction of the slurry viscosity.

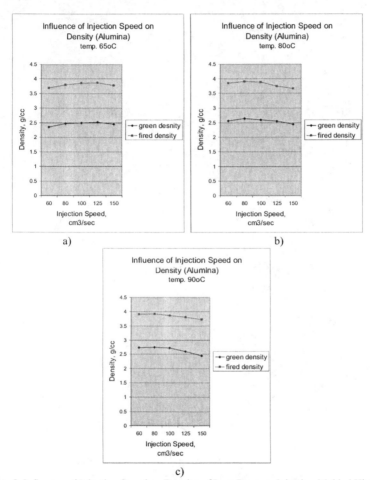

a)   b)

c)

Fig. 5. Influence of Injection Speed on Density of Low Pressure Injection Molded High Alumina Ceramics (firing temperature 1650°C) at Different Temperatures of Slurry (65-80-90°C). Process conditions: temperature of mold 15°C

The increase of temperature of the thermoplastic slurries positively affects their flowability and filling the mould cavity, especially in the case of the molds for complex shape components. When temperature of slurries increases from 60 to 90°C, less porosity (i.e. better compaction) in the ceramics is observed; shrinkage is reduced accordingly. However, at temperatures of 100°C or higher, paraffin wax starts evaporating, i.e. these high processing temperatures are not recommended. The influence of temperature of ceramic slurry on density is shown on Fig. 6.

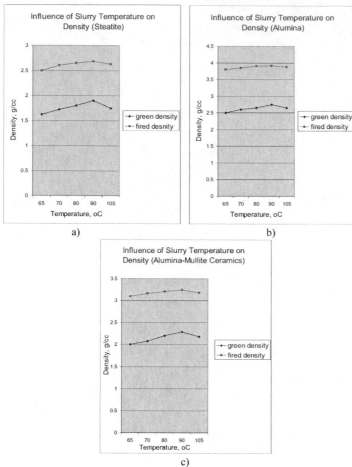

a)   b)

c)

Fig. 6. Influence of Temperature of Thermoplastic Slurry on Density of Low Pressure Injection Molded Ceramics
(a – steatite, firing temperature 1280°C; b – high alumina ceramics, firing temperature 1650°C; c – alumina-mullite ceramics, firing temperature 1410°C)
Process conditions: injection pressure 0.5 MPa, temperature of mold 15°C

Temperature of molds and difference between temperatures of slurry and molds are also important. The hardening of the bodies occurs "by layers"; the hardening starts from the mold surface that is obviously colder than the slurry. When the mold temperature is rather high and the difference in temperatures between slurry and mold is low, hardening occurs slowly with deformation of the injected body with possible hardening of the "whole" body. In this case, elevated porosity may be observed. In some cases, sticking to the mold surface may also occur. But when the mold is rather cold (10-15°C) and the difference in temperatures between slurry and mold is sufficient (greater than 50°C), the hardening occurs as "layer-by-layer" with minimal porosity and without deformation. In this case, the feeding of molds occurs without difficulties with no fast hardening in the mold feeding (sprue) area of the mold and in the middle of the body and with no residual cavity in the middle of the body. However, if the processing ceramic body has a complex shape with a necessity to use complex molds with many components, the temperature of the mold has to be elevated (e.g. 15-20°C). This also reduces the possible mechanical stresses occurred on the surface of the ceramic body.

The cooling of the molds, e.g. direction of the cooling, during injection molding process also needs to be conducted based on the particular shape of the ceramic components and mold design. The difference in setting and hardening of the injected bodies in the molds depending on the temperature gradients between the slurry and the mold may be explained by the features of crystallization of paraffin. If the difference in temperatures of the slurry and the mold is small (slow hardening), large paraffin crystals occur due to its migration from the interior not very solid areas of the injected body to its surface. As a result, the exterior body layers may have higher amounts of paraffin then the interior layers. But if the difference in temperatures of the slurry and the mold is sufficient, small paraffin crystals occur due to fast hardening. In this case, migration of paraffin occurs slowly, and the injected body has more even paraffin distribution, and, as a result, has higher density and mechanical strength. The influence of the temperature gradient (difference between slurry temperature and mold temperature) on density of alumina-mullite ceramics (as example) is shown on Fig. 7.

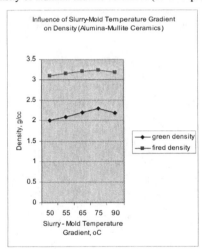

Fig 7. Influence of temperature gradient (slurry temperature – mold temperature) of thermoplastic slurry on density of low pressure injection molded alumina-mullite ceramics (firing temperature 1410°C) Process conditions: injection pressure 0.6 MPa, temperature of mold 15°C
Binder System Removal (Debinding)

The debinding of the low-pressure injection molded bodies is one of the key process operation steps. The features of this step are dealt with reversible melting of paraffin wax and similar organic materials. Therefore, the binder removal has to be conducted safely with elimination of the molded component warpage and destruction dealt with softening, melting, decomposition and evaporation of the used binder system. In order to avoid these negative factors, the debinding is usually conducted in the absorbent, i.e. the molded bodies are immersed into a ceramic powder placed into refractory saggers, and this powder promotes the safe binder removal absorbing molten organics via capillary forces during heating. The absorbent also supports the ceramic components while paraffin-based binder melting and prevents deformation of the components. This technique was described[8-11, 30-33] and optimized in industrial conditions. Debinding without powder absorbent is possible only for the small components with a height up to 10 mm and with a wall thickness up to 5 mm placed on the porous setters, which may absorb molten organics. In order to conduct debinding without absorbent, the content of the thermoplastic binder in the ceramic composition is desirable to be less than 10%. In any case, because of the "heavy" organic components have to be removed, the debinding process must be maintained in the furnaces with a strong exhaust system.

Thermoplastic binders, particularly, industrial paraffins, usually contain some amounts of oily components (these oily substances are used for the preparation of industrial paraffins). The processes occurring with the binder components transformation are utilized at the design of the debinding temperature profile. This process has a few steps, which are utilized in industrial manufacturing:

- At the heating of paraffin binders, the oily components start migrating out at temperatures of 50-60°C (so-named "sweating"). This step is conducted slowly (5-10°C/hr). At this condition, the ceramic components gain additional mechanical strength.
- The next heating step to temperatures of 100-150°C is also conducted slowly (5-10°C/hr) when paraffin binder intensively melts and moves into the absorbent and starts decomposing. At the end of this step, a soak may be successfully used.
- Molten paraffin decomposes intensively in the temperature range of 150-300°C and then finally evaporates. In the end of this stage, the level of decomposition (pyrolysis) of paraffin achieves 90-95%, and usually slow heating and sufficient soak at the end of this stage are used.
- The stage of the final decomposition of residual hydrocarbons in the middle of the body and their burning occur at the temperature range of 300-400°C; at this temperature stage, the debinding can be conducted at faster ramps of 10-20°C/hr.
- The final debinding step is continued up to 850-1150°C for strengthening of the green body. The final debinding temperature depends on the type of ceramics; for example, for BaTiO3 ceramics this temperature is ~900°C, steatite ceramics need 870-950°C for the final debinding, while oxide and alumina ceramics require 1050-1100°C.
- Cooling of the components after debinding is conducted rather fast because the ceramics has rather high porosity, but it is strong enough to withstand thermal and mechanical loads.

The debinding profile depends on the size and shape of the ceramic components; in some cases, additional temperature soak is applied during the debinding. The migration of liquid binder through the pore channels from the interior region to the surface occurs by the capillary action, and this migration depends on the particle size distribution and pore size. The ceramic components made from finer powders, which have smaller pore sizes (that is important to attain a high level of densification), demand a longer debinding cycle. It is also obvious that when larger size ceramic components are produced, longer debinding cycle has to be applied. Relatively fast heating, especially at the 1st-2nd stages mentioned above usually results in formation of the "hard skin" (described by J.E. Zorzi et al[31]) deteriorating the debinding body and resulting in internal stresses and crack appearance and deformation. Fast heating at the 3rd step also results in deformation of components and small blisters formation. The products after debinding when the described approach is applied, are quite mechanically strong despite their high porosity, and the removal of the absorbent can be conducted without difficulties.

Another important factor dealt with the necessity to use slow debinding process is dealt with the feature that, due to the softening and migration of the thermoplastic binder components, the ceramic particles also start migrating and rearranging their position. Slow debinding process promotes better particle rearrangement and compaction at this migration. The importance of the effect of particles mobility on densification was pointed out by Liu et al[32] studied the debinding of high pressure injection molded components (paraffin wax and vinyl acetate polymer were used as the major and secondary binder system components), and this feature is inherent to the processing of low pressure injection molded ceramics.

Safe debinding process (no cracks and other defects appearance) is possible not only when the slow debinding profile utilized the features, which occur with the organic substances at the heating, is used, but also when other optimal process parameters are maintained. Additionally to the already mentioned important processing factors, it should be noted that cracks and bubbles may appear in the cases of incorrect slurry formulation, e.g. when the surfactant contents are greater than optimal[19, 32], from particle flocculation during low-temperature reheating of not well stabilized slurry[34], when the high-melting point binder components with excessive amounts are used[33].

The final firing of the ceramic products is carried out using the "traditional" firing conditions and temperature profiles depending of the requirements for the particular type of ceramics. For example, fully dense low-pressure injection molded ceramics with properties comparable with properties of slip cast or isostatically-pressed ceramics are easy obtained. It is clear that higher fired density can be achieved in the case of better compaction of the ceramic particles in the green body, and it can be obtained in the case of higher solid contents in the thermoplastic slurries (of course, when all particles are wet by the liquid phase, and no voids exist in a green body).

EQUIPMENT FOR LOW PRESSURE INJECTION MOLDING

Equipment for low pressure injection molding includes an electrically heated tank for the binder system preparation equipped with a stirring device and an attached vacuuming system and a unit for casting under pressure (injection) applying compressed air. One of the leading manufacturers of this equipment is Peltsman Corp. (USA). This equipment can work in semi- and full-automation regimes to increase the productivity. A principle schematic of the slurry handling and injection molding device is shown in Fig. 2, and a general view of the machines is shown on Fig 8. A double blade planetary mixer with variable speed of rotation during mixing of the hot slurry in the heated tank and a vacuum pump for deairing of the slurry during mixing provide a high homogeneity of thermoplastic slurries. The principle of design of the injection molding machines allows to apply bottom injection using compressed air. The equipment is supplied with computerized instrumentations in order to install and monitor steering speed, temperature and pressure in the mixer, as well as in the transferring tube, injection pressure and time. The heated and de-aired slurry is injected under the applied pressure into the cold mold positioned on the top of the cover plate. The technology and equipment provide a versatile processing from the laboratory level to mass production. Due to automation in the forming process, a capacity of this equipment and principle of work allow to produce a few hundred small pieces/ day.

Metallic molds used for this technology are designed based on the required ceramic component shapes and dimensions, as well as utilizing the features of the technology, particularly, thermoplastic slurries behavior. The designed molds, equipment and technology allow to produce components with complex shapes, including the components with asymmetric parts, with the holes and details positioned not at the major axis directions. For the mass-production, multi-cavity molds are often used if the shape of the ceramic components is rather simple and the component design is suitable. In many cases, the molds consist of many parts; some molds may have ejector that increases manufacturing productivity. Also, the molds may have a device to cut the flash off that also promotes productivity. As mentioned above, the mold assembling-casting (injection)-demolding cycle may take from less than 1 min. to several min., i.e. productivity is high, and it may be increased in a few times in the case of the use of multi-cavity molds.

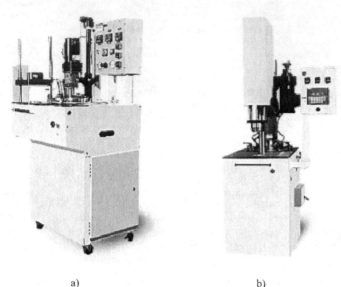

a)   b)

Fig 8. A General View of the Low Pressure Injection Molding Semi-Automatic Machine MIGL-33 (a) and Automatic Machine MIGL-37 (b) (Peltsman Corp.)

Because of low pressures and temperatures are used for the forming of ceramic components, low cost molds (e.g. steels, aluminum, brass) can be successfully used. Additionally to metallic molds, W. Bauer et al[13] mentioned about a possibility of silicone rubber molds for the forming. Due to the "soft" process conditions, the molds have very low wear, and they can work for several thousands forming cycles without degradation. In the case of a proper mold working surface preparation, a high level of the surface quality comparable with pressing and slip casting technologies may be attained. As mentioned above, no expensive diamond machining after firing is required for the near-net shape components.

Fig 9. Ceramic Components Produced by Low Pressure Injection Molding Technology

SUMMARY

Low pressure injection molding technology is a versatile manufacturing method that allows for the production of complex shape ceramic bodies in high-output quantities. The examples of different ceramic components are shown on Fig 9. Optimizing the process conditions and thermoplastic slurry composition for particular types of ceramics and product configurations, high quality ceramics can be produced. Among different factors affecting processing yield and quality of ceramics, the importance of the ceramic powder preparation, thermoplastic binder composition, thermoplastic slurry injection features and debinding process are outlined providing some major recommendation principles for industrial process.

REFERENCES:

[1] R.M. German, Powder Injection Molding, Metal Powder Industries Federation, Princeton, NJ, 1990

[2] R.M. German, S.-T.P. Lin, Key Issues in Powder Injection Molding, *Amer. Ceram. Soc. Bull.,* 1991, 70 (8) 1294-1302

[3] M.J. Edirisinghe, J.R.G. Evans, Review: Fabrication of Engineering Ceramics by Injection Molding. I. Materials Selection, *Int. J. High. Technol. Cer.,* 2, 1986, 1-31; II. Techniques, *Int. J. High. Technol. Cer.,* 2, 1986, 249-278

[4] M.J. Edirisinghe, Fabrication of Engineering Ceramics by Injection Molding, *Amer. Ceram. Soc. Bull.,* 1991, 70 (5) 824-828

[5] B.C. Mutsuddy, R.G. Ford, Ceramic Injection Molding, Chapman and Hall, UK, 1995

[6] T.J. Whalen, C.F. Johnson, Injection Molding of Ceramics, *Amer. Ceram. Soc. Bull.,* 1981, 60 (2) 216-220

[7] J.A. Mangels, R.M. Williams, Injection Molding Ceramics to High Green Densities, *Amer. Ceram. Soc. Bull.,* 1983, 62 (5) 601-606

[8] P.O. Gribovsky, Hot Casting of Ceramic Products, GosEnergoIzdat, Moscow-Leningrad, 1961 (in Russian)

[9] G.N. Maslennikova, F.Y. Kharitonov, N.S. Kostyukov et al, Processing of Electroceramics, Energia, Moscow, 1974 (in Russian)

[10] V.L. Balkevich, Technical Ceramics, Stroiizdat, Moscow, 1984 (in Russian)

[11] N.S. Kostyukov, E.Y. Medvedovski, F.Y. Kharitonov, Electroinsulating Corundum-Mullite Ceramic Materials, Far-East Division of Academy of Science USSR, Vladivostok, 1988 (in Russian)

[12] R. Knitter, W. Bauer, D. Gohring et al, Manufacturing of Ceramic Microcomponents by a Rapid Prototyping Chain, *Adv. Eng. Mat.,* 2001. 3 (1-2) 49-54

[13] W. Bauer, R. Knitter, Development of a Rapid Prototyping Process Chain for the Production of Ceramic Microcomponents, *J. Mater. Sci.,* 2002, 37, 3127-3140

[14] I. Krindges, R. Andreola, C.A. Perottoni et al, Low-Pressure Injection Molding of Ceramic Springs, *Int. J. Appl. Ceram. Technol.,* 2008, 5 (3) 243-248

[15] S. Novak, K. Vidovic, M. Sajko, T. Kosmac, Surface Modification of Alumina Powder for LPIM, *J. Eur. Cer. Soc.,* 1997, 17, 217-223

[16] S.T. Lin, R.M. German, Interaction between Binder and Powder in Injection Moulding of Alumina, *J. of Mater. Sci.,* 1994, 29, 5207-5212

[17] R.M. German, Homogeneity Effects on Feedstock Viscosity in Powder Injection Moulding, *J. Amer. Ceram. Soc.,* 1994, 77 (1) 283-285

[18] M.J. Edirisinghe, The Effect of Processing Additives on the Properties of a Ceramic-Polymer Formulation, *Ceram. Int.,* 1991, 17, 89-96

[19] T.-Y. Chan, S.-T. Lin, Effects of Stearic Acid on the Injection Molding of Alumina, *J. Amer. Ceram. Soc.,* 1995, 78 (10) 2746-2752

[20]D.-M. Liu, Effect of Dispersants on the Rheological Behavior of Zirconia-Wax Suspensions, *J. Amer. Ceram. Soc.*, 1999, 82 (5) 1162-1168

[21]J.E. Zorzi, C.A. Perottoni, J.A.H. Da Jornada, Wax-Based Binder for Low-Pressure Injection Molding and the Robust Production of Ceramic Parts, *Ind. Ceram.*, 2003, 23 (1) 47-49

[22]R. Lenk, A. Ph. Krivoshchepov, Effect of Surface-Active Substances on the Rheological Properties of Silicon Carbide Suspensions in Paraffin, *J. Amer. Ceram. Soc.*, 2000, 83 (2) 273-276

[23]M.J. Edirisinghe, J.R.G. Evans, Systematic Development of the Ceramic Injection Moulding Process, *Mater. Sci. & Eng.*, 1989, A109, 17-26

[24]W.R. Russel, D.A. Saville, W.R. Schowalter, Colloidal Dispersions, pp. 456-506, Cambridge University Press, Cambridge, UK, 1989

[25]S.M. Wolfrum, J.J. Ponjee, Surface Modification of Powders with Carboxylic Acids, *J. Mater. Sci. Lett.*, 1989, 8, 667-669

[26]J.H. Song, J.R.G. Evans, Ultrafine Ceramic Powder Injection Moulding: The Role of Dispersants, *J. Rheol.*, 1996, 40 (1) 131-152

[27]A. Dakskobler, T. Kosmac, The Effect of Interparticle Interactions of the Rheological Properties of Paraffin-Wax Suspensions, p. 213-220 in Advanced Processing and Manufacturing Technologies for Structural and Multifunctional Materials III (Ed. T. Ohji and M. Singh), Ceramics Engineering and Science Proceedings, 2009, v. 30, n. 8

[28]T. Kitano, T. Kataoko, T. Shirota, An Empirical Equation of the Relative Viscosity of Polymer Melts Filled with Various Inorganic Fillers, *Rheol. Acta*, 1981, 20, 207-209

[29]A.B. Metzner, Rheology of Suspensions in Polymer Liquids, *J. Rheol.*, 1985, 29, 739-775

[30]J.K. Wright, J.R.G. Evans, Removal of Organic Vehicle from Moulded Ceramic Bodies by Capillary Action, *Ceram. Int.*, 1991, 17, 79-87

[31]J.E. Zorzi, C.A. Perottoni, J.A.H. Da Jornada, Hard-skin development during binder removal from Al2O3-based green ceramic bodies, *J. Mater. Sci.*, 2002, 37 (9) 1801-1807

[32]D.-M. Liu, W.J. Tseng, Influence of Debinding Rate, Solid Loading and Binder Formulation on the Green Microstructure and Sintering Behaviour of Ceramic Injection Mouldings, *Ceram. Int.*, 1998, 24, 471-481

[33]W.J. Tseng, C.-K. Hsu, Cracking Defect and Porosity Evolution during Thermal Debinding in Ceramic Injection Moldings, *Ceram. Int.*, 1999, 25, 461-466

[34]J.H. Song, J.R.G. Evans, Flocculation after Injection Molding in Ceramic Suspensions, *J. Mater. Res.*, 1994, 9 (9) 22-26

CERAMIC INJECTION MOLDING USING A PARTIALLY WATER-SOLUBLE BINDER
SYSTEM: EFFECT OF BACK-BONE POLYMERS ON THE PROCESS

OxanaWeber[1, 2, 3] and Thomas Hanemann[1, 2]

[1] Karlsruhe Institute of Technology, Institute for Applied Materials/ Hermann-von-Helmholtz-Platz 1,
D-76344 Eggenstein - Leopoldshafen, Germany

[2] University of Freiburg, Department of Microsystems Engineering, Georges-Köhler-Allee 102, D-
79110 Freiburg, Germany

[3] Corresponding author: oxana.weber@kit.edu

ABSTRACT

The represented research work deals with a study concerning the process stability if different
raw materials of poly-(methyl-methacrylate) (PMMA) are applied as a back-bone polymer. Feedstocks
based on zirconia powder were observed, and polyethylene glycol was used as basic polymer. Three
kinds of commercially available PMMAs were compared which feature a similar molar mass
distribution and differ in their apparent condition only: granular, pearly and edged shaped. The results
of this study showed which essential influence this outward property of PMMA has on compounding
of homogeneous feedstocks, separation effects between powder and binder components, dimensional
stability of injection moulded and sintered parts as well as achieved densities of final micro products.

INTRODUCTION

With the established replication technique powder injection moulding (PIM) it is possible to
manufacture near-net-shape complex micro geometries cost effectively and multiplicatively [1]. The
schematic process of this technology is shown in Figure 1 and mainly consists of four steps: Preparing
of a homogeneous mix from organic binders and ceramic or metal powder particles also called
feedstock, shaping to a green body via injection moulding, removal of binder by chemical or thermal
treatment and subsequent sintering of a brown body to its final dense shape.

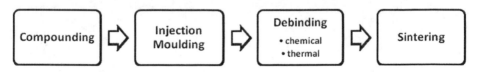

Figure 1: Schematic process of micro powder injection moulding.

Quality and stability of each process step as well as properties of final parts strongly depend on
the choice of a suited combination of binder polymers. In this research work only an environmental
friendly binder system will be considered based on water-soluble polyethylene glycol (PEG), whereby
in contrast to the conventional binders based on polyethylene/wax [2-5], their one advantage consists in
performing of the solvent pre-debinding in water and not in an organic substance such as hexane [4-5]
which is toxic, high flammable and hazardous. There are already many publications which cover the
combination of polyethylene glycol (PEG) and poly-(methyl-methacrylate) (PMMA) as a binder

system for powder injection moulding [6-8]. These research groups have proved that PEG/PMMA polymer matrix composites can be successfully used for production of highly dense metal components by means of PIM-technique. For this reason this study is concerned with the contrasting juxtaposing of three commercially available kinds of PMMA similar in chemical properties but strongly differ in their apparent condition in view of their lab as well as industrial handling and application.

EXPERIMENTAL MATERIALS AND EQUIPMENTS:

As an appropriate powder material for these experiments the submicron zirconia powder (TZ-3YS-E) from the Japanese Tosoh Corporation was chosen because of its excellent sintering properties. The powder is available in a spray-dried form as 30-50 μm large granulate [5] whereby the primary particles feature a close particle size distribution with a mean value of 440 nm [5]. The specific surface amounts to (6.6 ± 0.1) m^2/g [5].

A multi-component binder system was observed mainly consisting of a low viscous and water-soluble PEG from Carl Roth GmbH. In addition stearic acid was applied as surfactant in order to prevent the agglomeration of powder particles and to support interaction between the powder surface and binders during compounding guaranteeing a homogeneous feedstock. Therefore the effective amount of surfactant is usually related to the specific surface of the used powder. In this study the stearic acid concentration was set to a fixed value of 4.4 mg/m^2. Three kinds of PMMA were tested and compared for their suitability for PIM-process. Afterwards the materials will be shortened in the following way: PMMA SA from Sigma Aldrich, PMMA QA from Quinn Plastics and PMMA DG from Evonik/Degussa AG. The process relevant material data are summarized in Table I. If the glass transition temperature and the beginning of the decomposition are considered it is noticeable that the values measured by DSC-analysis are close to each other and typical for PMMA. In addition to the polymer characteristics given in Table I Figure 2 shows the relative molecular weight distribution obtained by size exclusion chromatography whereby the x-axes are pictured as a molar mass calibrated on polystyrol. It is obvious that in contrast to PMMA SA and PMMA QA which do not differ from each other the maximum of the distribution function of PMMA DG is shifted to the right by about 20,000 Dalton and moreover features a slightly wider width that is also reflected by the larger value of the polydispersity factor.

Table I: Characteristics of the used PMMA kinds. The table shows measured values.

	PMMA SA	PMMA QA	PMMA DG
Vendor	Sigma-Aldrich	Quinn Plastics	Degussa AG
Average molecular weight (10^3 g/mol)	52.4	55.3	74.7
Polydispersity factor	374.1	321.4	578.9
Glass transition temperature (°C)	103	115	103
Decomposition temperature (°C)	260	300	250

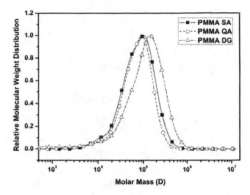

Figure 2: Results of size exclusion chromatography: Molar weight distribution of three PMMAs studied in this work.

Figure 3 shows the photographic images of the studied raw materials for back-bone polymer function. They differ strongly in their apparent condition: While PMMA QA features a fine, pearly-like spherical geometry close to perfection (Figure 3, middle) the morphology of PMMA SA is edged with a marked anisotropy and consists of extremely small crystallites as well as big informal chunks (Figure 3, left). In opposite to this, PMMA DG consists of larger, almost equally big pellets resulting from its production (Figure 3, right).

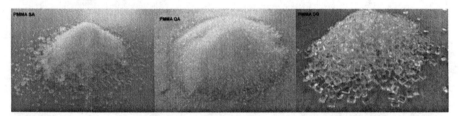

Figure 3: Photographic images of source substances for a back-bone polymer, from left to right: PMMA Sigma Aldrich (SA), PMMA Quinn Acryl (QA) and PMMA Degussa Granulate (DG).

After mixing of organic binder components with the ceramic TZ-3YS-E powder in a mixer kneader machine from Brabender W 50 EHT at 160°C and one hour kneading time the prepared feedstocks are rheologically characterized using a high pressure capillary rheometer from Goettfert Werkstoffpruefmaschinen GmbH. For a better comparison of compounding results and flow properties the viscosity measurements were performed at 160 °C, too. All viscosity graphs presented here are calculated by using Weissenberg-Rabinowitsch correction.

The tests on an injection moulding machine (Battenfeld Microsystem 50) were conducted using an isothermal process control. For proving the basic process ability of the prepared compounds a simple geometry in form of a disc was chosen. The dimensions of this injection moulded part had a diameter of 10.6 mm and a maximal thickness of one millimetre. In Figure 4 a 3D-CAD-model of a sample is given together with the gating system.

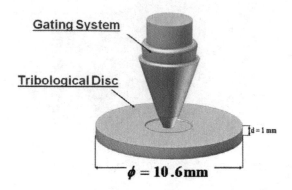

Gating System

Tribological Disc

d = 1 mm

$\phi = 10.6 \text{mm}$

Figure 4: Specimen geometry for the injection moulding tests [9].

For further treatment in order to get a dense and defect-free sintered product the organic constituent parts have to be removed again. Therefore the moulded green bodies underwent a two step debinding. At first the major part of PEG was extracted using deionised water for at least eight hours. Then the remaining polymer materials were thermally eliminated. The parameter for thermal debinding and subsequent sintering program can be found in Table II. The two last steps were carried out in the chamber furnace RHF17/3E from Carbolite. The thermal decomposition of PMMA starts at a temperature around 250°C under oxygen atmosphere. Therefore it is essential to apply a low heating rate to avoid deformations in the final parts.

Table II: Thermal debinding and sintering parameters for ceramic feedstocks based on PMMA binder system.

Step		Temperature (°C)	Heating/ Cooling rate (°C/min)	Duration (min)
Debinding	Heating up	250	0.5	
	Dwell	250		30
	Heating up	320	0.5	
	Dwell	320		30
	Heating up	380	0.5	
	Dwell	380		30
	Heating up	500	2	
	Dwell	500		30
Sintering	Heating up	1450	5	
	Dwell	1450		60
	Cooling down	Room temperature	5	

RESULTS AND DISCUSSION:

To compare the compounding behaviour of feedstocks based on different PMMAs the TZ-3YS-E powder was kneaded together with polymeric mix of PMMA, PEG and stearic acid. The volume portions of the binder compound and the zirconia divide one to one. Figure 5 shows the typical course of the kneading curves consists of three zones. The first zone is the filling zone and lasts from circa 5 till 7 minutes. In the kneading zone the powder agglomerates are broken and the powder surface is wetted by organic polymers. This section is characterized by an extremum of torque and a subsequent drop of the measure curve which ends in the steady zone where the constant grade of dispersion is achieved and the torque does not change anymore [5, 10]. It is important to keep the width of the kneading zone small to prevent a heat generation by the friction of the powder particles which can lead to oxidation and associated degradation of polymers. As torque measurements showed that all three PMMA resulted a close kneading region whereby in contrast to the other two back-bone polymers only PMMA SA featured a obviously higher maximum by a factor 1.5 caused by cracking of big informal pellets. It also needs to be said that the kneading zone of PMMA DG was shifted to the right by about 5 minutes because its rough granulate had to be pre-melted in the kneading chamber for preventing of unneeded height torques and hence for a better mixing with the powder as well as the other binder components. All three compounds achieved the steady zone very fast and the same value of the final torque after one hour kneading time which provides information about the homogeneity of the prepared feedstocks.

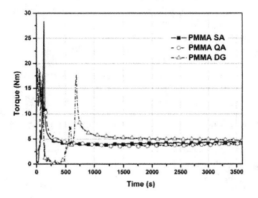

Figure 5: Compounding behaviour of zirconia feedstocks depending on the kind of the back-bone polymer.

The effect of binder compounding on the flow characteristics of the PMMA based feedstocks with the same powder loading of 50 Vol.% TZ-3YS-E was investigated, too. Results of rheological evaluation are presented in Figure 6. All graphs obviously exhibit pseudo plastic flow behaviour which is characterized by decreasing of viscosity with shear rate. In the whole measurement range the viscosity gradually increases from PMMA QA above PMMA DG to PMMA SA by a factor of (1.3±0.2) respectively. Despite of the equal molecular weight distribution of PMMA SA and QA the feedstock consisting of PMMA QA could be characterised by a better homogeneity and sufficiently lower viscosity for injection moulding which also allows a possible further rising of filler loading. In contrast due to the slightly higher average molecular weight of PMMA DG one would expect a viscosity increase. But the measured curve is located between PMMA AS and PMMA QA. This can be explained by the higher polydispersity factor which means a broader molecular weight distribution and shorter polymer chains which can act as a lubricant and lower the polymers viscosity.

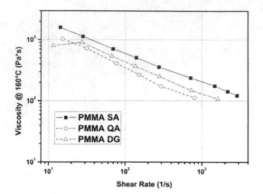

Figure 6: Influence of PMMA on the feedstock viscosity at constant zirconia load of 50 Vol.%.

The injection moulding of the polymer/powder composite to a green body is a very important and critical step in the whole process. Due to the multiphase character of the feedstock and extremely high shear rates occurring during the shaping separations between powder particles and binder components even in a highly homogenous feedstock can be generated. This leads to an agglomeration of particles as well as to a particle density variation in the moulded parts hence causing anisotropic shrinkage during the sintering step. It was possible to produce the simple discs of all feedstocks considered here in some cases even till to 52.5 Vol.% of zirconia. Table III shows the general differences in the moulding characteristics between three kinds of PMMA. While an automatic mode could be achieved for compounds consisting of 50 Vol.% ceramic powder and PMMA QA or DG as back-bone polymer it was necessary to increase the filler load up to 52.5 Vol.% for feedstocks based on PMMA SA due to sticking of the gate system on the nozzle side. A further mentionable peculiarity was the strong occurrence of abrasions on the surface of moulded parts starting from the injection point if PMMA SA was applied. This phenomenon can be explained by the higher feedstock viscosity as described earlier. Also after sintering this kind of circular discoloration was observed (Figure 7) and could be verified as ferric deposition on the sample surface by using XPS analysis.

Table III: Overview about moulding filling characteristics.

Solid Load (Vol. %)	PMMA SA	PMMA AS	PMMA DG
50	gate sticks, abrasions	automatic mode	automatic mode
52.5	automatic mode, lower abrasions	irregular metering	-

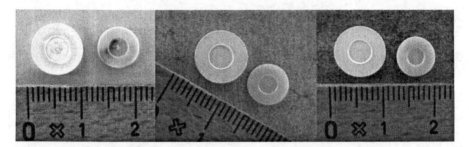

Figure 7: Green and sintered body of tribological discs resulted from compounds based on 50 Vol.% TZ-3YS-E powder, the back-bone polymer used from left to right: PMMA Sigma Aldrich, PMMA Quinn Acryl and PMMA Degussa Granulate.

The densities of green and sintered parts are listed in Table IV. It was possible to produce final products without any deformation or cracks. The achieved densities of all shaped PIM-samples investigated in this work showed a good accordance with the literature density value if zirconia of 6.10 g/cm^3.

Table IV: Density of moulded and sintered test specimens prepared by feedstocks filled with 50 Vol.% of zirconia as measured by helium pycnometry.

	Density (g/cm^3)	
Kind of PMMA	Green body	Sintered part
SA	3.62	6.08
QA	3.61	6.03
DG	3.57	6.07

CONCLUSION:

The results presented in this article consider three commercially available kinds of poly-(methyl-methacrylate) which are similar in their chemical properties such as molecular mass distribution or glass transition temperature but strongly differ in their apparent conditions: edged shaped, pearly or granulate. The suitability of these back-bone polymers was proven in all PIM process steps in view of their lab as well as industrial handling and application. Summarized it can be said, that mainly polymers with a close molecular weight distribution as well as fine spherical geometry with a high specific surface are most suitable for micro PIM. Because these properties ensure a fast melting of organic polymers during compounding which results in a homogeneous mix of powder particles and binder matrix, a high flow ability for injection moulding and low separation effects during shaping. General properties of these studied PMMAs and a weighted comparison of them are summarized and listed as an overview in Table V.

Table V: Weighted comparison of three differ back-bone polymers. The grade of comparison means: A -top/ height potential, B - good/ middle potential; C - sufficient/low potential.

Properties	PMMA SA	PMMA QA	PMMA DG
Compounding/ Lab	A	A	B
Compounding/ Industry	A	A	C
Injection moulding	B	B	B
Homogeneity	B	A	C
Strength of green body	A	A	A
Contour accuracy/ abrasion	C	A	A

ACKNOWLEDGEMENTS:

The authors thank the Deutsche Forschungsgemeinschaft (DFG) for financial support within the Collaborative Research Centre SFB 499 and all colleagues at KIT Campus North involved in this work: Especially, K. Georgieva for feedstock preparation and characterisation, P. Holzer for sample fabrication, Dr. S. Schlabach and I. Fuchs for determination of densities by helium pycnometry, Dr. M. Tosoni and O. Schwindt for helpful support in carrying out of the SEC-measurements as well as Dr. M. Bruns for performing of the XPS-investigations on the sintered specimen.

REFERENCES:

[1] R.M. German and A. Bose, Injection moulding of metals and ceramics, Metal Powder Industries Federation, Princeton, New Jersey U.S.A. (1997).
[2] V. Piotter, W. Bauer, T. Hanemann, M. Heckele and C. Mueller, Replication technologies for HARM devices: status and perspective, Microsystems Technology, Volume 14 (2008), Pages 1599–1605.
[3] V. Piotter, G. Finnah, B. Zeep, R. Ruprecht, and J. Hausselt, Metal and ceramic micro components made by powder injection moulding, Materials Science Forum, Volumes 534-536 (2007), Pages 373-376.
[4] A. Ruh, T. Hanemann, R. Heldele, V. Piotter, H.-J. Ritzhaupt-Kleissel and J. Hausselt, Development of Two-Component Micropowder Injection Molding (2C MicroPIM): Characteristics of Applicable Materials, International Journal of Applied Ceramic Technology, Volume 8, Number. 1 (2011), Pages 194-202.
[5] R. Heldele, Entwicklung und Charakterisierung von Formmassen für das Mikropulverspritzgießen, Doctoral Thesis, University of Freiburg, Freiburg/Breisgau (2008).
[6] N. Chuankrerkkul, H.A. Davies and P.F. Messer, Application of PEG/PMMA binder for powder injection moulding of hard metals, Materials Science Forum, Volume 561-565 (2007) Pages 953-956.
[7] S. Eroglu and H.I. Bakan, Solvent debinding kinetics and sintered properties of injection moulded 316L stainless steel powder, Powder Metallurgy, Volume 48, Number 4 (2005), Pages 329-332.
[8] N. Chuankrerkkul, P.F. Messer and H.A. Davies, Flow and void formation in powder injection moulding feedstocks made with PEG/PMMA binders Part 1 – Experimental observations, Powder Metallurgy, Volume 51, Number 1 (2008), Pages 66-71.
[9] O. Weber, T. Mueller and T. Hanemann, Development of polymer matrix composites for the realisation of ceramic and metal microcomponents using micro powder injection moulding, 14th European conference on composite materials, Hungary/ Budapest (2010), Paper ID: 899-ECCM14.

[10]   O. Weber, A. Loges and T. Hanemann, Injection moulding of sub-micrometer zirconia powder using partially water-soluble organic binder, proceeding of the 12th Conference of the European Ceramic Society - ECerR XII, Sweden/Stockholm (2011).

GREEN-CONSCIOUS CERAMIC INJECTION MOLDING

OxanaWeber[1, 2, 3], Thomas Hanemann[1, 2]

[1] Karlsruhe Institute of Technology, Institute for Applied Materials/ Hermann-von-Helmholtz-Platz 1, D-76344 Eggenstein - Leopoldshafen, Germany

[2] University of Freiburg, Department of Microsystems Engineering, Georges-Köhler-Allee 102, D-79110 Freiburg, Germany

[3] Corresponding author: oxana.weber@kit.edu

ABSTRACT

The present work focuses on the development of an alternative binder system for the application in micro ceramic injection moulding consisting of environmentally friendly polyethylene glycol (PEG) and polyvinyl butyral (PVB). In comparison to conventional binders based on wax this new combination has an important profit that the liquid pre-debinding step takes place in water and not in a toxic organic solvent such as hexane. In addition, these compounds allow further process simplifications by its economic and timesaving benefits. The results presented here cover all process steps from feedstock preparation to sintering of final micro parts. In comparison to literature significantly higher filler loadings with zirconia powder (TZ-3YS-E) of 55 vol% were achieved with and reproducible injection moulding. The solvent debinding experiments showed that the best dimensional stability was found at room temperature where about 94% of the PEG-content was extracted within 2 hours. Furthermore, removal of polymers can be performed just thermally without any deformation. A concluding sintering at 1450°C led to final defect-free products with a high density close to 100%.

INTRODUCTION

Powder injection moulding (PIM) provides an opportunity to produce complexly shaped ceramic micro components with high sintered density [1-3]. The advantage of this technique is most of all the possibility to manufacture in high numbers of pieces with near-net-shape geometries without any post-processing steps. The process consists of four major steps: compounding of a mixture of fine powder and polymeric binders to a feedstock, shaping to a green body via injection moulding, then removal of the binder by using a chemical solvent or thermally and following sintering of the brown body to the final dense product. The several process steps have been extensively studied and are described in the literature in detail [1].

Binder systems based on polyethylene-wax-mix are quite established and already well-studied in the high pressure injection moulding [3-6]. Apart from many advantages such as high green strength and versatile usability these binders have an important disadvantage because the chemical pre-debinding of the wax component takes place in organic solvent such as hexane or heptane [4] which is toxic and environmentally questionable. As an alternative water-soluble polymers can be applied. The most known representative of this group is polyethylene glycol (PEG) which is often used in medicine and pharmacy. This research work shows that this material in combination with high molecular polyvinyl butyral (PVB) which is responsible for a sufficient strength of the moulded micro parts can be also successively deployed in the technical field as a binder devise. To prove the aptitude of this new binder system for production of micro ceramic parts all process steps above mentioned were surveyed and are presented in this report.

EXPERIMENTAL MATERIALS AND EQUIPMENTS:

In general the application of the final product defines the choice of the suitable powder materials. In the manufacturing of complexly shaped micro components the major attention is paid to powders with small particle size which should be at least about one order of magnitude smaller than the minimal dimension of the finished micro parts [1]. Figure 1 shows a scanning electron microscopy (SEM) image of zirconia powder (TZ-3YS-E, Tosoh Corporation, Japan) used for this work presented here. The powder consists of sub-micrometer big primary particles with an edged morphology which tend to formation of large agglomerates of 30-50 μm. Further important properties of this filler material are summarised in Table I.

Figure 1: SEM image of primary particles of used powder: TZ-3YS-E.

Table I: Characteristics of the used TZ-3YS-E zirconia powder. The table shows measured values expected the theoretical density.

Solid Material	Vendor	Type	Average particle size d_{50} (μm)	Spec. surface area (m^2/g)	Theo. density (g/cm^3)
ZrO_2	Tosoh Corp.	TZ-3YS-E	0.44[7]	6.60[7]	6.10

As already mentioned a multi-component binder system was applied to produce injection moulded components: The major fraction also called basic polymer is comprised of low viscous PEG ®Rotipuran from Carl Roth GmbH. High molecular PVB from Kuraray Europe GmbH was used as a back-bone polymer to provide enhanced stability of green specimen during demoulding and the solvent debinding step in water. Three types of PVB (PVB T, PVB SP and PVB SH) were tested here whereby they differ in the average molecular weight from PVB T to PVB SH by a factor of two. The overview about relevant characteristics of used base and back-bone polymers is given in Table II. The next unit of the applied binder system was stearic acid (Carl Roth GmbH, Germany) added as surfactant which acts as intermediary agent between powder surface and binder matrix and helps so to achieve a better distribution of ceramic particles and polymers in order to produce a homogeneous feedstock. The surfactant concentration was calculated on the specific surface of the powder amount used in the current experiment.

Table II: Overview about important characteristics of two major binder components PEG and PVB.

	Base - polymer	Back – Bone - polymer
Name	Polyethylene glycol (PEG)l	Polyvinyl butyral (PVB)
Molecular formula	$C_{2n}H_{4n+2}O_{n+1}$	$H_2(C_8H_{14}O_2)_n$
Vendor	Carl Roth GmbH	Kuraray Europe GmbH
Glass transition temperature (°C)	35-62	63-75
Decomposition temperature (°C)	280	220
Solubility	Water	Alcohols, glycol ether
Commercial application	Pharmaceutical and cosmetic industry, biology, medicine	Varnishes, adhesives, films, food packaging industry

All binder components were mixed with zirconia powder using a mixer kneader machine (Brabender W 50 EHT, Germany). The compounding tests were performed at 125°C, where mixing rotation frequency was set for 30 rpm and the compounding time lasted 60 min. Figure 2 shows an exemplarily torque curve as a function of kneading time, which typically consists of three zone:

i. Zone 1 is the filling zone where the torque increases in succession of the charge of the kneading chamber with powder and binder components;
ii. Zone 2 is the kneading zone. Here the powder surface is wetted by binder polymers and powder agglomerates are gradually broken. The consequence is a homogenisation of the powder-binder-mix and hence the decreasing of the kneading torque;
iii. In the zone 3 (steady zone) the torque does not change anymore if a constant grade of dispersion and wetting is achieved. The value of the final torque gives a possibility for estimation of the homogeneity of the prepared feedstock [7, 8].

To investigate the influence of the solid load and the surfactant concentration on the flow ability of the compounds the rheological properties of the feedstocks were measured at 160°C using a high pressure capillary melt rheometer from Goettfert Werkstoffpruefmaschinen GmbH, Germany. As a standard measurements setup a capillary with a feeding angle of 180°, a total length of 30 mm and a hole diameter of 1 mm was used. All viscosity curves measured were mathematically corrected by Weissenberg-Rabinowitsch formula [9].

To determine the suitability of the new compounds for injection moulding a Microsystem 50 machine from Battenfeld GmbH, Austria was used. Three different geometries were chosen for the shaping experiments: tribological disc, flexural micro part and micro gear housing. All shaping tests were carried out using an isothermal process control.

After moulding the binder should be removed again. For that reason the green bodies were chemical debinded in de-ionized water at different temperatures for maximal 16 hours with continually stirring of solvent. In this step the major binder amount, namely PEG, was extracted which simplifies the following thermal treatment where the remaining binder components were burned out. For these experiments a chamber furnace RHF17/3E (Carbolite GmbH, Germany) was used. In Table III the temperature program applied for thermal debinding and sintering steps is specified in detail.

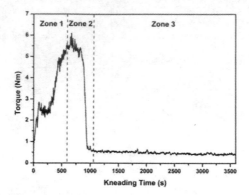

Figure 2: Typical behaviour of kneading torque during compounding of a feedstock. Three zones can be defined: Filling (1), kneading (2) and steady zone (3).

Table III: Debinding and sintering parameters for ceramic feedstocks based on PEG/PVB binder system.

Step		Temperature (°C)	Heating/ Cooling rate (°C/min)	Duration (min)
Debinding	Heating up	220	0.5	
	Dwell	220		30
	Heating up	350	0.5	
	Dwell	350		30
	Heating up	450	0.5	
	Dwell	450		30
	Heating up	500	2	
	Dwell	500		30
Sintering	Heating up	1450	5	
	Dwell	1450		60
	Cooling down	Room temperature	5	

As a quality control feature the estimation of the density and the light microscopy were taken. The density of green bodies was determinate by using the Helium pycnometry (Phycnomatic ATC from ThermoFinigan/Porotec). In case of the sintered final parts the Archimedes method was applied.

RESULTS AND DISCUSSION:

In order to make a decision on which PVB can be used as back-bone polymer with sufficient homogeneity and flow ability of the ceramic feedstocks best, the three kinds of PVBs were compared concerning the compounding behaviour. The feedstocks were prepared using 50 Vol.% of TZ-3YS-E powder and 4.4 mg/m^2 stearic acid as surfactant. In Figure 3 the final torque after feedstock kneading is shown as a function of the PVB-kind. The increase of molar mass of the back-bone polymer causes a marked rise of the final torque. By addition of PEG ®Rotipuran the torque values was reduced in all three cases by factor (3.0±0.1). That indicates a better plasticization and homogenisation of the ceramic compounds and makes the feedstock flowable for injection moulding. For this reason the next experiments concentrated on the PVB T/ PEG ®Rotipuran based binder systems only.

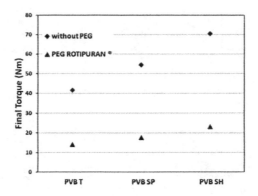

Figure 3: Influence of the binder composition on the final kneading torque. The considered compounds were prepared with 50 Vol. % TZ-3YS-E and 4.4 mg/m^2 stearic acid as surfactant.

The next step was to find the maximal processable solid load in the zirconia compounds. The results of the so-called study of filler loading carried out are presented in Figure 4 and showed a typical increase of the torque with the rising powder amount. If the critical solid concentration is achieved the torque falls off, because a satisfactory wetting of the full powder surface by binder polymers was not ensured anymore and the remaining powder particles are not transported by the mixing hooks. In this case the resulted torque corresponds to a value of a respectively less filled compound. As one can discern from the diagram (Figure 4) for feedstocks observed here the critical filler loading of 56 Vol.% TZ-3YS-E was determined.

The effect of the surfactant concentration between 1.1 and 8.8 mg/m^2 on the rheological behaviour of 55 Vol.% filled zirconia feedstocks can be seen in Figure 5. The given viscosity was measured at 160°C and taken at the shear rate of 100 1/s. In the range between 1.1 and 3.3 mg/m^2 of stearic acid ratio the principle of operation of a surface-active substance is clear to see. The decreasing viscosity indicates a better distribution of powder particles in the binder matrix and hence an improving homogeneity of prepared compounds. From 3.3 to 4.4 mg/m^2 the viscosity features a constant plateau what can be defined as an optimal concentration. Compared to conventional polyethylene-wax based binders [7] the same behaviour of stearic acid can be found. This proves that the necessary amount of surfactant substance is to be calculated related to the specific surface area of powder only, no matter whether a polar or a nonpolar binder system is used. A further addition of stearic acid may reduce the viscosity but in this case the surfactant is abundantly and fills a role of a lubricant. The consequences are a lower concentration of the back-bone polymer PVB and hence an insufficient stability of the moulded micro parts.

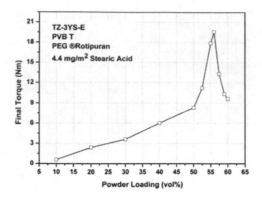

Figure 4: Study of filler loading of feedstock based on PVB T and PEG ®Rotipuran. The stearic acid concentration was set to a fixed value of 4.4 mg/m^2.

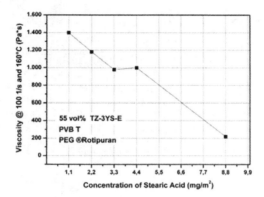

Figure 5: Influence of the stearic acid concentration on the rheological properties of the 55 Vol.% filled ceramic feedstocks.

Shaping tests via an injection moulding machine showed that feedstock were reproducibly processed under isothermal conditions up to 55 Vol.% referring to the specimen geometry. The results are summarized in Table IV as an overview. Compared to the established wax-polyethylene binders the loading value for TZ-3YS-E usually amounts to 40 till maximal 50 Vol. % [2, 4 and 7]. Thereby this eco-friendly binder combination gives a possibility to improve the dimensional stability by reduced sintering shrinkage.

After successful moulding to desired geometries the specimen underwent chemical treatment in water for removal of the main part of PEG component. The used experimental set ups were described above. Figure 6 represents exemplarily the debinding behaviour of tribological discs compounded with 50 Vol.% of TZ-3YS-E and 4.4 mg/m^2 stearic acid. Results showed the relative mass loss of PEG

®Rotipuran based on the initial content depending on extracted temperature and storage time in water. After only 2 hours the extracted amount of PEG overcomes 90 % and then approximates to a constant value. The influence of solvent temperature is more likely to be observed in the green part stability than in the mass loss during the debinding step. Some marked deformations such as blister and cracks were detected at higher temperatures and became larger in the range between 40 to 50°C (Figure 7). In contrast specimen treated at room temperature featured defect-free shape.

Table IV: Results of injection moulding tests.

Tested geometries	Process control	Reproducible up to
Tribological disc	Isothermal	52.5 Vol.% TZ-3YS-E
Flexural micro part	Isothermal	55 Vol.%TZ-3YS-E
Gear housing	Isothermal	52.5 Vol.%TZ-3YS-E

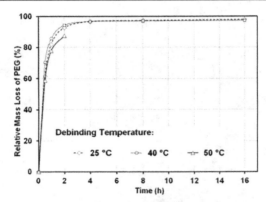

Figure 6: Results of chemical pre-debinding step in water of 50 Vol.% filled ceramic discs as a function of immersion time and solvent temperature.

Figures 8 and 9 show two further examples of tested geometries: A gear housing and a flexural micro part as green (left) and sintered (right) body were realized by 55 Vol.% filled zirconia compounds. The thermal debinding program is listed in Table III. In both cases it was possible to produce final products without any deformations or cracks whereby the density achieved came to (6.10 ±0.05) g/cm³ which effectively corresponds to hundred percentage of the theoretical density of zirconia.

Figure 7: Photographic image of a at 50°C solvent debinded tribological disc: Cracks formation across the whole area.

Figure 8: Micro gear housing: Green (left) and under air condition sintered body (right) realized with 55 Vol.% zirconia feedstock.

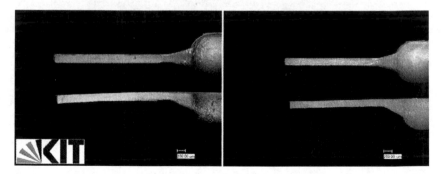

Figure 9: Flexural micro part: Green (left) and under air condition sintered body (right) realized with 55 Vol.% zirconia feedstock.

CONCLUSION:

For production of highly dense ceramic micro components or specimen with structural micro details using micro powder injection moulding ecological friendly polymer matrix composites based on combination of water soluble PEG as base component and PVB as back-bone polymer were surveyed. This research work reports about each process step in detail: In comparison to conventional on wax based binder systems described in literature [2, 4 and 7] significantly higher solid portion up to 55 Vol.% of TZ-3YS-E were reproducibly compounded to homogeneous feedstocks and successful processed to stable green bodies. This can help to improve dimensional stability of final parts by reduced sintering shrinkage. Further process simplifications allowed by PEG/PVB combination is a fast and harmless removal of PEG in water at room temperature without any deformations. A subsequent sintering at 1450°C led to final defect-free products with a high density close to 100%. Future research efforts will mainly concentrate on testing of this binder system with other specimen geometries in micro as well as in macro injection moulding.

ACKNOWLEDGEMENTS

The authors thank the Deutsche Forschungsgemeinschaft (DFG) for financial support within the Collaborative Research Centre SFB 499 and all colleagues at KIT Campus North involved in this work: Especially, A. Loges and K. Georgieva for feedstock preparation and characterisation, P. Holzer for sample fabrication, Dr. S. Schlabach and I. Fuchs for determination of densities by helium pycnometry.

REFERENCES

[1]   R.M. German and A. Bose, Injection moulding of metals and ceramics, Metal Powder Industries Federation, Princeton, New Jersey U.S.A. (1997).

[2]   V. Piotter, W. Bauer, T. Hanemann, M. Heckele and C. Mueller, Replication technologies for HARM devices: status and perspective, Microsystems Technology, Volume 14 (2008), Pages 1599–1605.

[3]   V. Piotter, G. Finnah, B. Zeep, R. Ruprecht, and J. Hausselt, Metal and ceramic micro components made by powder injection moulding, Materials Science Forum, Volumes 534-536 (2007), Pages 373-376.

[4]   A. Ruh, T. Hanemann, R. Heldele, V. Piotter, H.-J. Ritzhaupt-Kleissel and J. Hausselt, Development of Two-Component Micropowder Injection Molding (2C MicroPIM): Characteristics of Applicable Materials, International Journal of Applied Ceramic Technology, Volume 8, Number. 1 (2011), Pages 194-202.

[5]   T. Hanemann, R. Heldele, T. Mueller and J. Hausselt, Influence of stearic acid concentration on the processing of ZrO2-containing feedstocks suitable for micropowder injection molding International Journal of Applied Ceramic Technology, Volume 8 (2011), Pages 865-872.

[6]   T. Hanemann, R. Heldele, and J. Haußelt, Structureproperty relationship of dispersants used in ceramic feedstock development, Proceedings 4M 2007 – 3rd International Conference on Multi-Material-Micro-Manufacture (4M), Borovets, Bulgarien (2007) Pages 73-76.

[7]   R. Heldele, Entwicklung und Charakterisierung von Formmassen für das Mikropulverspritzgießen, Doctoral Thesis, University of Freiburg, Freiburg/Breisgau (2008).

[8]   O. Weber, A. Loges and T. Hanemann, Injection moulding of sub-micrometer zirconia powder using partially water-soluble organic binder, Proceeding of the 12th conference of the European Ceramic Society - ECerR XII, Sweden/Stockholm (2011).

[9]   A.A. Collye and D.W. Clegg, Rheological measurement, Elsevier Applied Science, London (1998).

SHAPING OF LARGE-SIZED SPUTTERING TARGETS

Alfred Kaiser
LAEIS GmbH
Wecker, Luxembourg

ABSTRACT

Sputtering targets are used in PVD coating plants (PVD = physical vapor deposition) as a source for the coating material. Materials to be used can be indium tin oxide (ITO), alumina doped zinc oxide (AZO) and others. Large plates of various sputtering materials with an area of up to more than 0.5 m² and a thickness of < 10 mm up to approx. 40 mm are produced using uniaxial hydraulic presses. Despite their high aspect ratio the plates show good green strength and can be handled without problems. The uniaxial pressing is used either as the only shaping method for a subsequent pressureless sintering or as pre-densification for a second compaction step in an isostatic press. The advantage there is reduction of scrap to a minimum and to increase the throughput capacity of the isostatic press.

INTRODUCTION

"Sputtering" means a PVD technique (PVD = physical vapor deposition) used to produce thin films coatings[1]. Atoms of the source material, the target, are ejected in a high vacuum chamber by energetic particle bombardment, usually an argon plasma, and are deposited onto the surface of a substrate which is also placed in the vacuum chamber. Target materials can be different kind of metals and alloys, but also oxide or non-oxide ceramic materials like TiO_2, ZrO_2, TiN, ZnO, Al_2O_3, AZO (alumina doped zinc oxide), ITO (indium tin oxide) and many others. They are used to create functional coatings, e.g. with well-defined mechanical, optical and/or electrical properties. Substrates are displays, touchscreens, photovoltaic cells electronic components, silicon wafers and many others[2].

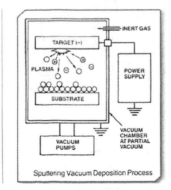

Figure 1.   Sputtering process principle
(picture courtesy of TCB Thermal Conductive Bonding, Inc.)

In order to achieve a sufficient homogeneity of the deposited film, the target should be larger than the substrate. Since the dimensions of the substrates to be coated are ever increasing, also large-sized sputtering targets are required. The shaping of sputtering targets is often done using an isostatic pressing process. With increasing target dimensions, the necessary isostatic presses become very expensive and, due to their comparably long cycle time, become a bottleneck in production capacity.

The company Laeis in Luxembourg, being a renowned supplier of uniaxial hydraulic presses amongst others for advanced ceramic products, introduced uniaxial hydraulic presses for the production of large-scaled sputtering targets several years ago. In the meanwhile, a full range of presses of different sizes have been supplied to the related industry. The intention of using such presses is to realize a precompaction for a subsequent isostatic pressing, thus making better use of the capacity of the isostatic press and to reduce the amount of scrap material due to the near net shaping capability of the uniaxial press. But there is also an increasing number of applications, where uniaxial pressing is used as the only shaping technology, followed by a pressureless firing process.

UNIAXIAL HYDRAULIC PRESSING TECHNOLOGY

When thinking about a suitable shaping technology for a special product, different selection criteria have to be taken into account. First of all there are criteria coming from product requirements like product geometry (length, width, height, aspect ratio, simple or complex shapes, undercuts etc.), special product requirements (e.g. high density, even density distribution, microstructure, contamination restrictions) or material properties (densification behaviour, sintering behaviour, reactivity against moisture and possible additives). Then we have process related influences like required throughput capacity, necessity of near net shape production, possibility of additional downstream processing steps like machining, grinding, polishing, etc.). All these criteria can reduce the number of available shaping technologies in industrial production plants significantly, compared to available laboratory scale shaping technologies which might have been applied during the R & D phase. Furthermore, it has to be checked, whether a selected technology has been proven already for similar application, whether there are different suppliers in the market who can provide for technical assistance and support, maintenance, spare parts etc. Last but not least commercial aspects like capital expenditure and operating costs (CAPEX / OPEX) and also sustainability play an important role.

Uniaxial hydraulic pressing is always one of the shaping technologies to be considered, because it is a well proven technology which has been used in practically all fields of ceramics production for many decades and has been improved continuously. It allows a comparatively „dry" pressing with low or even no moisture content of the mix body, which reduces the drying effort to a great extend, compared for example to casting or extrusion technologies. In many cases the drying step can be integrated into the firing process and no separate dryers are necessary. Products with high green density and good green strength can be manufactured despite a low binder content. Furthermore, low dimension tolerances, high contour sharpness and a good surface quality are advantages of this shaping technology, especially when compared to isostatic pressing or to low pressure casting. Optimized hydraulics, flexible control systems and a high degree of automation provide for a high throughput capacity and lead to a good economic efficiency. Even though uniaxial hydraulic pressing is an "old" technology which is widely used in the fields of traditional ceramics like tiles, refractories, etc. it is also a very up-to-date technology which conquered it´s share in advanced ceramics production. Small-sized parts like carrier rings for the textile industry, ferrites, piezoceramics, etc. for the automotive industry, electronics and many other applications are produced in huge numbers for a long time[3]. More recently, however, also a steadily growing number of advanced ceramic products of larger size are uniaxially pressed[4,5].

For shaping of large-sized sputtering targets with high aspect ratios and a thickness of typically 8 - 15 mm modified tile presses have proven to be very useful. Basically, such presses compact the material only from one side (unilateral pressing), since differences in density from top to bottom are tolerably small for thin products. To improve the density distribution, the mould frame can be designed movable. During the compaction cycle it is moved downwards by material friction, thus causing a partial densification force also to be applied from bottom [figure 2].

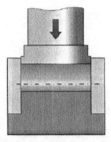 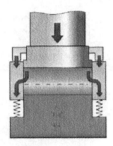

Figure 2.   Unilateral pressing principles (compaction only from top side)
  left: basic principle with fixed lower die and fixed mould frame
  right: advanced principle, mould frame movable, cushioned by
            mechanical spring or hydraulically

If products with larger thickness are required, bilateral pressing becomes more favourable. Two principle variations are shown in figure 3. Details of the various pressing principles are described elsewhere [6,7] and are not discussed here.

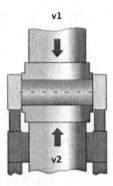 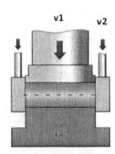

Figure 3.   Bilateral pressing principles (compaction from top and bottom side)
  left: basic principle with movable upper and lower die and fixed mould frame
  right: advanced HPF principle with movable upper die, fixed lower die and movable
            mould frame.
  If $v_1 = v_2$ on the left side and $v_2 = 0.5 * v_1$ on the right side, the same densification
  behaviour will result.

Everybody who ever tried to press advanced ceramic products using a standard hydraulic press, knows about the difficulties and problems related to de-airing during compaction and re-expansion during ejection. Such effects can hardly be influenced by standard press control systems. LAEIS control technology provides for the necessary features to compact the parts and remove them from the mould without defects. With such a very flexible control system numerous parameters can be adjusted in a wide range to the requirements of a given task. For instance it is possible to vary the movement of the mould filling box (speed, reversed movements, "shaking") in order to get a homogeneous filling of the mould cavity also when the material shows a poor flowability. During the pressing cycle, parameters like punch entry speed, pressure increase curve, pressure holding time, ejection speed, load during ejection etc. are further parameters which can be exactly regulated according to the material's densification behaviour. Closed loop control of all axes compensates external influences, guarantees a precise approach of the setpoints with a very accurate reproducibility and thus helps to obtain highest geometrical precision while at the same time the cycle time can be reduced[8].

De-aeration strokes and/or mould evacuation remove part of the entrapped air from the mould cavities and contribute to a higher green density as well as avoid structural inhomogeneities in the pressed body. Especially the vacuum pressing technology has widened the range of applications of uniaxial hydraulic pressing to a great extend[9] and is more and more becoming a standard feature for challenging pressing tasks. Typical examples are large-sized sputtering targets, where the air has to travel a long distance until it can escape via the narrow mould gap. This often leads to the formation of cracks or layers in the plates which is particularly annoying when such failures become evident only in the subsequent firing process. Evacuation of the mould to a predefined degree before starting of the compaction eliminates that problem.

SPUTTERING TARGETS: APPLICATION EXAMPLES

During the last years, LAEIS GmbH successfully introduced uniaxial hydraulic presses to producers of sputtering targets. The range of available presses for this application is shown in table I.

Table I. Technical data of LAEIS hydraulic presses available for the production of sputtering targets.

press type	pressing force		useful die area	maximum filling depth	maximum ejection force
	kN	t	mm x mm	mm	kN
Alpha 800	8,000	800	1,170 x 500	60	140
Alpha 800/120	8,000	800	1,170 x 500	120	300
Alpha 1500	15,000	1,500	1,320 x 640	60	200
Alpha 1500/120	15,000	1,500	1,320 x 640	120	300
Alpha 4200	42,000	4,200	1,600 x 1,100	60	280
Omega 3000	30,000	3,000	1,600 x 1,000	60	280
PH 6500[*]	64,000	6,400	2,275 x 990	60	150

[*] in cooperation with Sacmi/Imola

In close collaboration with the end users of the presses, suitable pressing parameters have been defined for several target materials, including ZnO, AZO, ITO and others. Target sizes ranging from 150 x 150 mm² up to more than 2,000 x 400 mm² (> 0.8 m²) are being produced. The thickness of the targets is typically between 8 and 15 mm (in green state), but also plates up to 25 mm and more can be pressed. In some cases the uniaxial pressing is followed by an additional isostatic pressing step, but in other cases the targets directly undergo a debindering and firing process. In the first case the precompaction in the uniaxial press leads to a better utilization of the capacity of the isostatic press and also to a remarkable reduction of scrap material. This is a very important fact, especially when extremely expensive materials like ITO are used.

In some cases the technical center of LAEIS GmbH investigated the complete process chain starting from raw materials sourcing through recipe optimization, process optimization, laboratory and production scale tests up to toll production for all production steps, as exemplified in figure 4 for AZO sputtering targets.

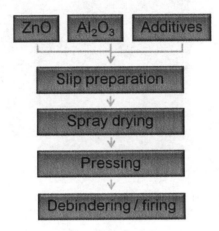

Figure 4.   Production steps in the complete process chain for AZO sputtering targets manufacture

On the left side of figure 5 the filled mould (300 x 400 mm²) of a press Alpha 1500/120 is shown, than the pressed plate after ejection (middle). The picture on the right gives an impression of the high green strength of the plates. They can be taken from the press and handled manually without any problems.

Figure 5.   Pressing of AZO sputtering targets (300 x 400 mm², thickness approx. 20 mm) in a press type Alpha 1500/120.
       left: mould cavity filled with spray dried powder
       middle: pressed plate after ejection from the mould
       right: manual handling of a green plate immediately after pressing

Figure 6 shows pictures of various presses used for sputtering targets production with a picture of a very large sputtering target pressed in the press type PH 6500 in figure 7. A comprehensive overview of the press types already in use for this application is given in table II together with an indication of approximate target sizes made in these presses.

Figure 6.   Various presses used for the production of sputtering targets.
left: Alpha 1500/120
middle: Omega 3000
right: PH 6500

Figure 7.   Large-sized sputtering target (approx. 1,500 x 350 mm²), pressed in a press type PH 6500.

Table II. LAEIS uniaxial hydraulic presses supplied for production of sputtering targets.

press type	pressing force		target material	length	width	height	typical applied pressure	
	kN	t		mm	mm	mm	kN/cm²	MPa
Alpha 1500/120	15,000	1,500	AZO, ZnO	430	320	≤ 35	6 - 10	60 - 100
			HA, AZO, ITO, ZnO, ...	400	300	5 - 30	5 - 12	50 - 120
Alpha 4200	42,000	4,200	HA, AZO, ITO	680	600	8 - 14	10	100
				1,600	430	≈ 10	6	60
Omega 2100	21,000	2,100	ITO	1,000	360	8	< 6	< 60
				500	250	8 - 12	≤ 15	≤ 150
Omega 3000	30,000	3,000	ITO	900	400	8 - 14	8	80
PH 6500 *)	64,000	6,400	ITO	1,700	400	8 - 15	8	80
				2,200	400	8 - 15	< 7	< 70

*) in cooperation with Sacmi/Imola

SUMMARY
        Sputtering targets of various composition (ITO, AZO, HA, etc.) are being pressed using uniaxial hydraulic pressing technology. Depending on press size and optimal selection of pressing parameters, very large target sizes of up to 0.8 m² with target thicknesses of up to 35 mm can be realized. Uniaxial hydraulic pressing as a well proven shaping technology can be used as the only shaping technology, but also as pre-densification for subsequent isostatic pressing; making better use of isostatic press capacity and reducing the production of waste material. The technical center of LAEIS is ready to evaluate the optimal technology in cooperation with our customers and also to offer toll production.

REFERENCES
[1] R. Behrisch (ed.); Sputtering by Particle Bombardment, *Springer, Berlin (1981)*, *ISBN 978-3540105213*
[2] J.L. Vossen; Transparent Conducting Films, in *Physics of Thin Films, Vol. 9. Edited by G. Haas, M. H. Francombe and R. W. Hoffman, Academic Press, New York 1977, 1–71*
[3] L. Lackner; Multi-Axial CNC Pressing Technology, *cfi/Ber.DKG 84 (2007) [10] E16-E19*
[4] A. Kaiser, R. van Loo, J. Kraus, A. Hajduk; Comparison of Different Shaping Technologies for Advanced Ceramics Production, *cfi/Ber.DKG 86 (2009) [4] E41-E48*

[5] A. Kaiser, R. Lutz; Uniaxial Hydraulic Pressing as Shaping Technology for Advanced Ceramic Products of Larger Size, *Interceram 60 (2011) [3] 230-234*

[6] W. Schulle; Die Preßformgebung in der Keramik, *Keram. Z. 44 (1992) [11],754-759 (in German)*

[7] V. Ramakrishnan; Modern Developments in the Fabrication Process of High-Grade Refractory Bricks, *Sprechsaal 120 (1987), 288-296 + 880-885*

[8] R. Lutz; Use of Closed Loop Controls in Hydraulic Press Forming of Ceramic Products to Obtain Highest Dimensional Accuracy, *Proceedings of the International Colloquium on Refractories, Aachen (2004), 222-224*

[9] A. Kaiser, R. Kremer; Fast Acting Vacuum Device: Guaranteed Quality for Pressed Refractories, *Interceram Refractories Manual 2003, 28-33*

TEM OBSERVATION OF THE Ti INTERLAYER BETWEEN SiC SUBSTRATES DURING DIFFUSION BONDING

H. Tsuda[1], S. Mori[1], M. C. Halbig[2] and M. Singh[3]
[1]Graduate School of Engineering, Osaka Prefecture University, Osaka, Japan
[2]NASA Glenn Research Center, Cleveland, Ohio, USA
[3]Ohio Aerospace Institute, NASA Glenn Research Center, Cleveland, Ohio, USA

ABSTRACT

Diffusion bonding was carried out to join SiC to SiC substrates using titanium interlayers. In this study, 10 μm and 20 μm thick physical vapor deposited (PVD) Ti surface coatings, and 10 and 20 μm thick Ti foils were used. Diffusion bonding was performed at 1250°C for PVD Ti coatings and 1200°C for Ti foil. This study investigates the microstructures of the phases formed during diffusion bonding through TEM and selected-area diffraction analysis of a sample prepared with an FIB, which allows samples to be taken from the reacted area. In all samples, Ti_3SiC_2, $Ti_5Si_3C_x$ and $TiSi_2$ phases were identified. In addition, TiC and unknown phases also appeared in the samples in which Ti foils were used as interlayers. Furthermore, Ti_3SiC_2 phases show high concentration and $Ti_5Si_3C_x$ formed less when samples were processed at a higher temperature and thinner interlayer samples were used. It appears that the formation of microcracks is caused by the presence of intermediate phase $Ti_5Si_3C_x$, which has anisotropic thermal expansion, and by the presence of an unidentified Ti-Si-C ternary phase with relatively low Si content.

INTRODUCTON

Silicon carbide (SiC) is a very promising material for high-temperature structural and extreme environment applications due to its excellent high temperature mechanical properties, oxidation resistance and thermal stability. Therefore, SiC has been used in a wide range of practical applications, not only as a monolithic material, but also as matrix and fiber reinforcement materials in composites[1]. However, further practical applications of complex shaped SiC components are hindered by geometrical limitations of current fabrication methods (i.e. hot pressing and chemical vapor deposition). To overcome this problem, joining approaches are being developed as a solution to fabricating large, three dimensional components from simpler-shaped ceramics. Various joining methods are being developed using reaction bonding[2-4] and brazing[5-6].

In previous studies, Gottselig et al[7] and Naka et al[8] have reported the bonding of silicon carbide with titanium and phase reaction and diffusion path in the ternary system. However, processing conditions were different and the microstructural analysis was not presented in detail. In previous studies, we have reported the details of various phases that form in the bonded area during diffusion bonding of SiC to SiC using scanning electron microscopy (SEM), X-ray diffraction (XRD) analysis, and energy dispersive spectroscopy (EDS) as well as electron microprobe analysis (EMPA)[9-11].

In this study, diffusion bonding was utilized to join SiC to SiC using two types of Ti interlayers. The first was a PVD Ti coating on the surface of SiC substrates, and the other was thin metallic Ti foil. After diffusion bonding, the effects of interlayer type, thickness, and processing temperature were studied in detail by analyzing the microstructure and phase formation. This paper contains detailed microstructural analysis of the phases formed during diffusion bonding through TEM and selected-area diffraction (SAD) analysis of samples prepared with an FIB (focused ion beam), which allows samples to be taken from the reacted area.

EXPERIMENTAL

CVD β-SiC substrates were obtained from Rohm & Hass (Woburn, MA). One interlayer type used for joining was titanium metallic foil from Goodfellow Corporation (Glen Burnie, MD) with thicknesses of 10 and 20 μm. A second interlayer type consisted of titanium coatings with a thickness of 10 μm which were applied in-house employing physical vapor deposition (PVD). To obtain a 10 μm PVD Ti interlayer, one coated and one uncoated SiC substrate were matched. To obtain a 20 μm PVD Ti interlayer, two surface Ti coated SiC substrates were matched. Before joining, all materials were ultrasonically cleaned in acetone for 10 minutes. Joints formed using PVD Ti coated SiC were diffusion bonded at 1250°C with a clamping pressure of 24 MPa. Joints formed using 10 and 20 μm Ti foil as the interlayer were diffusion bonded at 1200°C with a clamping pressure of 30 MPa. All joint processing was conducted in a vacuum environment with a 2 hr hold at the peak temperature under load followed by a slow cool down at a cooling rate of 5°C per minute.

Elemental analysis and phase identification was conducted on carbon coated samples using EMPA (JEOL 8200 Super Probe) for the bonds formed with PVD Ti and using SEM (JEOL, JXA-8900) for bonds formed with the Ti foil. Both were coupled with EDS. All samples for TEM were prepared by FIB (FEI, Quant 3D). The preparation process is described in detail in the literature[12]. By preparing the TEM specimen using the FIB, a clean, less-damaged, and precisely selected thin specimen was obtained, in contrast to specimens obtained using other methods, such as ion milling. Thus, the FIB allows the preparation of a TEM sample from the diffusion bonded area. Transmission electron microscopy (TEM) was conducted at 200kV (JEOL, JEM-2000FX).

RESULTS AND DISCUSSION
SEM Microstructure of Diffusion Bonded Samples

In Figures 1 (a) and (b), EMPA back-scattered electron images of the diffusion bonds are shown for when 10 μm (sample 1) and 20 μm (sample 2) thick PVD Ti coatings were used as the interlayers, respectively. In Table 1, the compositions as determined by EDS are listed for all phases identified in the diffusion bonds of sample 1 and sample 2. Microcracks can be seen and there exist three phases in the bond for the thicker 20 μm layer (Fig. 1(b)). The phases labeled A2, B2, and C2 were identified as Ti_3SiC_2, $TiSi_2$, and $Ti_5Si_3C_x$, respectively. On the other hand, the bond formed with the thinner 10 μm layer, as seen in Fig. 1(a), did not have microcracking. The two phases of Ti_3SiC_2 (phase A1) and $TiSi_2$ (phase B1) were again identified. The phases identified are consistent with the results of Gottselig et al[7]. The $Ti_5Si_3C_x$ phase was not observed in the thinner bond. Naka et al[8] suggested that $Ti_5Si_3C_x$ is an intermediate phase that is not present when phase reactions have gone to completion.

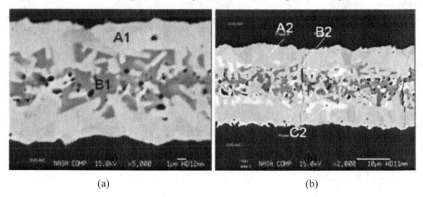

(a)   (b)

Figure 1. EMPA back-scattered electron images of diffusion bond of sample 1 (a) and 2 (b).

Table 1. Compositions of phases determined by EDS.

Sample	Phase Label	Composition in Atomic %			Probable phase
		C	Si	Ti	
Sample 1	A1	25	19	56	Ti$_3$SiC$_2$
	B1	3	61	36	TiSi$_2$
Sample 2	A2	26	18	56	Ti$_3$SiC$_2$
	B2	2	62	36	TiSi$_2$
	C2	7	34	59	Ti$_5$Si$_3$C$_x$

Figures 2(a) and (b) show the diffusion bonds formed with 10 μm (sample 3) and 20 μm (sample 4) thick Ti foils and Table 2 gives the compositions of the phases identified in each joint. The two joints shared two phases, labeled A and B in Table 2 and in the corresponding micrographs. The composition for phase A was in the range of 51–55C/13–14Si/32–35Ti (referred to as common phase A) and that for phase B was 38–47C/22–27Si/31–35Ti (referred to as common phase B). In sample 3, microcracks were scarcely observed or only very minimal microcracks were observed. However, in sample 4, significant microcracking was observed and an example of a typical microcrack is shown in Figure 2 (b). The microcracking seemed to be caused by the presence of labeled C phase. This will be discussed later.

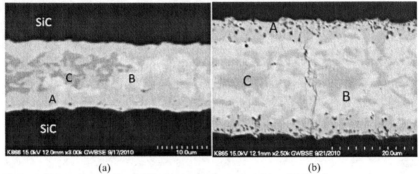

(a)   (b)

Figure 2. SEM micrographs of diffusion bond of sample 3 (a) and 4 (b).

Table 2. Compositions of phases determined by EDS.

Sample	Phase Label	Composition in Atomic %			Common and Unique Phases
		C	Si	Ti	
Sample 3	A	51	14	35	common phase A
	B	38	27	35	common phase B
	C	37	43	20	phase unique to this condition
Sample 4	A	55	13	32	common phase A
	B	47	22	31	common phase B
	C	58	7	35	common phase A with lower Si content

TEM of Diffusion Bonded Samples

TEM observations were conducted to examine in detail the microstructures of formed phases in the four diffusion bonded samples. Figure 3 shows the TEM micrograph of the sample in which 10 μm thick PVD Ti was used as the interlayer. It is clearly shown that the diffusion bond consists of many small reaction formed grains with length of 2~4μm, and width of 1~2μm. SAD patterns have been taken at locations which are shown in numerical order in the figure and the probable phases are summarized in Table 3. This type of detailed phase identification was conducted for all four samples with phases of about 30 grains in each sample determined from SAD pattern analysis (individual phases are distinguished by the differently shaped marks in the micrographs and are listed in corresponding tables). In addition, Table 4 gives the calculated percentage content of each phase. Fig. 4 shows SAD patterns of individual phases of sample 1. The TEM analysis indicates that sample 1 consists of three phases, namely Ti_3SiC_2, $Ti_5Si_3C_x$, and $TiSi_2$. Calculated fractions of phases formed during diffusion bonding are 91.4% Ti_3SiC_2, 2.9% $Ti_5Si_3C_x$, and 5.7% $TiSi_2$. Naka et al.[8] confirmed the presence of Ti_3SiC_2 and $TiSi_2$ in the final stage of processing by XRD when Ti was used to join SiC to SiC. As mentioned above, Ti_3SiC_2 is the primary phase and the fractions of other phases are extremely small. This suggests that the reaction between the SiC substrate and the 10 μm PVD Ti coating interlayer may have been complete in sample 1.

Table 3. Summary of probable phases determined from SAD patterns.

No	Sample1	Sample2	Sample3	Sample4	No	Sample 1	Sample2	Sample3	Sample4
1	$Ti_5Si_3C_x$	Ti_3SiC_2	Ti_3SiC_2	Ti_3SiC_2	19	Ti_3SiC_2	Ti_3SiC_2	Ti_3SiC_2	TiC
2	Ti_3SiC_2	Ti_3SiC_2	Ti_3SiC_2	Ti_3SiC_2	20	Ti_3SiC_2	Ti_3SiC_2	Ti_3SiC_2	Ti_3SiC_2
3	Ti_3SiC_2	$Ti_5Si_3C_x$	Ti_3SiC_2	Ti_3SiC_2, TiC	21	Ti_3SiC_2	$Ti_5Si_3C_x$	unknown	$Ti_5Si_3C_x$
4	Ti_3SiC_2	$Ti_5Si_3C_x$	Ti_3SiC_2	$Ti_5Si_3C_x$	22	Ti_3SiC_2	Ti_3SiC_2	unknown	$Ti_5Si_3C_x$
5	Ti_3SiC_2	Ti_3SiC_2	$Ti_5Si_3C_x$	$Ti_5Si_3C_x$	23	Ti_3SiC_2	$Ti_5Si_3C_x$	$Ti_5Si_3C_x$	$Ti_5Si_3C_x$
6	Ti_3SiC_2	Ti_3SiC_2	$Ti_5Si_3C_x$	Ti_3SiC_2	24	Ti_3SiC_2	Ti_3SiC_2	Ti_3SiC_2	$Ti_5Si_3C_x$
7	Ti_3SiC_2	Ti_3SiC_2	$Ti_5Si_3C_x$	Ti_3SiC_2	25	Ti_3SiC_2	Ti_3SiC_2	Ti_3SiC_2	Ti_3SiC_2
8	Ti_3SiC_2	Ti_3SiC_2	TiC	Ti_3SiC_2	26	Ti_3SiC_2	Ti_3SiC_2	Ti_3SiC_2	unknown
9	Ti_3SiC_2	Ti_3SiC_2	Ti_3SiC_2	$Ti_5Si_3C_x$	27	Ti_3SiC_2	Ti_3SiC_2	Ti_3SiC_2	$Ti_5Si_3C_x$
10	Ti_3SiC_2	Ti_3SiC_2	Ti_3SiC_2	$TiSi_2$	28	$TiSi2$	Ti_3SiC_2	Ti_3SiC_2	$Ti_5Si_3C_x$
11	Ti_3SiC_2	Ti_3SiC_2	$TiSi_2$	$Ti_5Si_3C_x$	29	Ti_3SiC_2	$TiSi_2$	Ti_3SiC_2	unknown
12	$TiSi_2$	Ti_3SiC_2	Ti_3SiC_2	$Ti_5Si_3C_x$	30	Ti_3SiC_2		Ti_3SiC_2	$Ti_5Si_3C_x$
13	Ti_3SiC_2	Ti_3SiC_2	Ti_3SiC_2	$Ti_5Si_3C_x$	31	Ti_3SiC_2		Ti_3SiC_2	Ti_3SiC_2
14	Ti_3SiC_2	Ti_3SiC_2	TiC	Ti_3SiC_2	32	Ti_3SiC_2		Ti_5Si_3Cx	
15	Ti_3SiC_2	$TiSi_2$	$TiSi_2$	$Ti_5Si_3C_x$	33	Ti_3SiC_2		Ti_5Si_3Cx	
16	Ti_3SiC_2	$TiSi_2$	Ti_3SiC_2	Ti_3SiC_2	34	Ti_3SiC_2			
17	Ti_3SiC_2	Ti_3SiC_2	Ti_3SiC_2	Ti_3SiC_2	35	Ti_3SiC_2			
18	Ti_3SiC_2	Ti_3SiC_2	Ti_3SiC_2	TiC					

Table 4. Calculated fractions of phases formed during diffusion bonding.

	Sample 1	Sample 2	Sample 3	Sample 4
Ti_3SiC_2	91.4	75.9	63.5	37.5
Ti_5Si_3Cx	2.9	13.8	18.2	43.8
TiSi2	5.7	10.3	6.1	3.1
TiC	0.0	0.0	6.1	9.1
unknown	0.0	0.0	6.1	6.3
Total	100	100	100	100

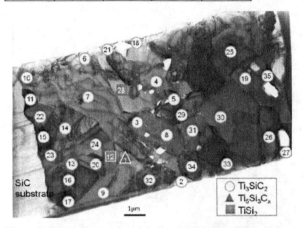

Figure 3. TEM micrograph and determined phases of sample 1.

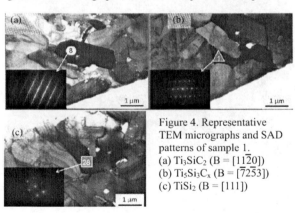

Figure 4. Representative TEM micrographs and SAD patterns of sample 1.
(a) Ti_3SiC_2 (B = [11$\overline{2}$0])
(b) $Ti_5Si_3C_x$ (B = [$\overline{7}$2$\overline{5}$3])
(c) $TiSi_2$ (B = [111])

Figure 5 is a TEM micrograph taken from sample 2 in which the determined phases are shown by the different marks. Ti_3SiC_2, $Ti_5Si_3C_x$, and $TiSi_2$ were again identified. Fractions of the phases are 75.9% Ti_3SiC_2, 13.8% $Ti_5Si_3C_x$, and 10.3% $TiSi_2$. It is noted that compared with the case for sample 1, the fraction of Ti_3SiC_2 was less and the fraction of $Ti_5Si_3C_x$ was considerably greater. The reactions between SiC and Ti seemed to be complete for a 2 hr hold when the thinner 10 μm Ti foil interlayer was used. In contrast, the reaction seemed to be incomplete when processed at the same temperature and for the same hold time when the thicker 20 μm Ti interlayer was used.

In Fig. 6, TEM micrograph and determined phases of sample 3 are shown. Ti_3SiC_2, $Ti_5Si_3C_x$, $TiSi_2$, and TiC were detected in sample 3. Moreover, there were also unknown phases. Fractions of the phases were 63.5% Ti_3SiC_2, 18.2% $Ti_5Si_3C_x$, 6.1% $TiSi_2$, 6.1% TiC, and 6.1% unknown phase.

Figure 7 shows TEM micrographs and observed phases in sample 4. The calculated fractions of each phase are 37.5% Ti_3SiC_2, 43.7% $Ti_5Si_3C_x$, 3.1% $TiSi_2$, 9.4% TiC, and 6.3% unknown phase. Also, in Fig. 8, representative TEM micrographs and SAD patterns obtained from sample 4 are shown.

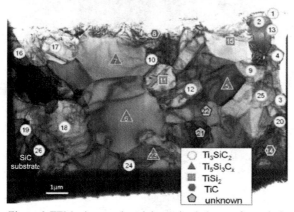

Figure 5. TEM micrograph and determined phases of sample 2.

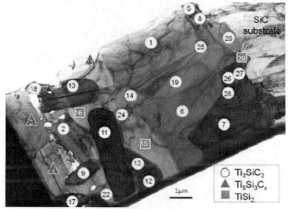

Figure 6. TEM micrograph and determined phases of sample 3.

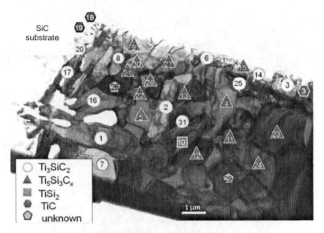

Figure 7. TEM micrograph and determined phases of sample 4.

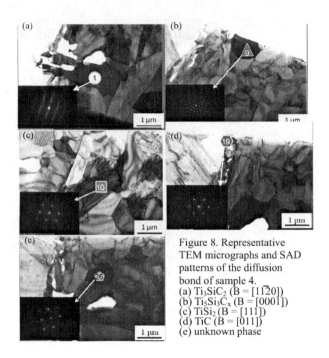

Figure 8. Representative TEM micrographs and SAD patterns of the diffusion bond of sample 4.
(a) Ti_3SiC_2 (B = [$11\bar{2}0$])
(b) $Ti_5Si_3C_x$ (B = [$000\bar{1}$])
(c) $TiSi_2$ (B = [111])
(d) TiC (B = [$0\bar{1}1$])
(e) unknown phase

Relation between Microstructure and Microcracking

The two sets of joints formed with the PVD Ti interlayer (thicknesses of 10 μm and 20 μm) and with the Ti foil interlayer (thicknesses of 10 μm and 20 μm) were formed during 2 hour holds at 1250°C and 1200°C, respectively. In the joint for the thinner PVD Ti interlayer, only stable phases were present. Thus, it is assumed that chemical reaction between SiC substrate and Ti interlayers had been completed. Thus sample 1 is superior with having no microcracks. While, in the joint with the thicker PVD Ti interlayer (Sample 2), a lot of the detrimental intermediate $Ti_5Si_3C_x$ grains were detected. According to the literature[8], it is known that the $Ti_5Si_3C_x$ has been recognized as a solid solution in which a small atomic percentage of carbon was contained in Ti_5Si_3. Also, this intermediate phase has anisotropic thermal expansion[13,14]. As the joint is cooled after processing, mismatches of the coefficient of thermal expansion may induce thermal stresses, and this would explain the reason for microcrack formation observed in sample 2,

On the other hand, the joint formed with the 10 μm Ti foil (sample 3) and holding for 2 hr at 1200°C had almost the same fraction of $Ti_5Si_3C_x$ observed in sample 2. However, in sample 3, a third phase (phase C) with composition 37C/43Si/20Ti was present. The relatively high Si content in this phase may help increase ductility so that stresses are alleviated and microcracks do not form. Therefore, microcracks were scarcely observed or only very minimal microcracks were observed in sample 3.

The joints formed with the 20 μm Ti foil (sample 4) and holding for 2 hr at 1200°C had a different third phase (phase C) with composition 58C/7Si/35Ti (referred to in Table 2 as "common phase A with lower Si content"). In this bond, phase C with low silicon content may be less ductile and therefore contribute to the observed microcracking. The extensive microcracking is mainly attributable to the high fraction of intermediate $Ti_5Si_3C_x$ phase. If the unknown phase detected in Sample 4 has content similar to that of "common phase A with lower Si content (labeled C)", this may also induce microcracking in sample 4.

CONCLUSION

Two types of Ti (PVD coated Ti on SiC substrate and metallic Ti foil) with different thicknesses of 10 and 20 μm were used to join SiC to SiC through diffusion bonding. Joints formed with the PVD Ti coated SiC were processed at 1250°C and joints formed with Ti foil as an interlayer were processed at 1200°C. After diffusion bonding, the microstructure of the bonded region was revealed by TEM. The results are summarized as follows.

(1) Ti_3SiC_2, $Ti_5Si_3C_x$ and $TiSi_2$ were identified in all samples. Furthermore, TiC and unknown phases appeared in the samples in which Ti foils were used as the interlayer (samples 3 and 4). Furthermore, Ti_3SiC_2 formed more and $Ti_5Si_3C_x$ formed less when samples were processed at a higher temperature and a thinner interlayer was used.

(2) In the sample with 10 μm PVD Ti interlayer (sample 1), the fraction of Ti_3SiC_2 was highest and the fraction of $Ti_5Si_3C_x$ was lowest or zero. Thus the sample was superior in terms of having minimal microcracks. On the other hand, in the sample with 20 μm PVD Ti interlayer (sample 2), less Ti_3SiC_2 and more $Ti_5Si_3C_x$ formed compared with the case for sample 1. Thus, a lot of microcracking was observed due to the presence of $Ti_5Si_3C_x$.

(3) Although, in the sample with 10μm Ti foil interlayer (sample 3), microcracks were scarcely observed. This may relate to the presence of a phase with relatively high Si content phase. Moreover, if the unknown phase detected in the sample with 20 μm Ti foil interlayer (sample 4) has content similar to that of phase with relatively lower Si content, this may also cause microcracking.

ACKNOWLEDGEMENT

H. T. would like to thank Mrs. Taeko Yuki and Mr. Tsukasa Koyama of Osaka Prefecture University for preparing TEM samples using an FIB.

REFERENCES
[1]P. J. Lamicq, G. A. Bernhart, M. M. Dauchier, and J. G. Mace, "SiC/SiC Composite Ceramics," Am. Ceram. Soc. Bull., 65 [2] (1986) 336–338.
[2]M. Singh, "A Reaction Forming Method for Joining of Silicon Carbide-based Ceramics," Scripta Materialia, Vol. 37, Issue 8 (1997) 1151-1154.
[3]M. Singh, "Joining of Sintered Silicon Carbide Ceramics for High Temperature Applications," Journal of Materials Science Letters, Vol. 17, Issue 6 (1998) 459-461.
[4]M. Singh, "Microstructure and Mechanical Properties of Reaction Formed Joints in Reaction Bonded Silicon Carbide Ceramics," Journal of Materials Science, 33 (1998) 1-7.
[5]V. Trehan, J.E. Indacochea, and M. Singh, "Silicon Carbide Brazing and Joint Characterization," J. Mech. Behavior of Materials, Vol. 10, Issue 5-6 (1999) 341-352.
[6]M.G. Nicholas, "Joining Processes: Introduction to Brazing and Diffusion Bonding," Springer (1998).
[7]B. Gottselig, E. Gyarmati, A. Naoumidis, and H. Nickel, "Joining of Ceramics Demonstrated by the Example of SiC/Ti," Journal of the European Ceramic Society, Vol. 6 (1990) 153-160.
[8]M. Naka, J. C. Feng, and J. C. Schuster, "Phase Reaction and Diffusion Path of the SiC/Ti System," Metallurgical and Materials Transactions A, Vol. 28A (1997) 1385-1390.
[9]M. Singh and M.C. Halbig, "Bonding and Integration of Silicon Carbide Based Materials for Multifunctional Applications," Key Engineering Materials, Vol. 352 (2007) 201-206.
[10]M.C. Halbig and M. Singh, "Development and Characterization of the Bonding and Integration Technologies Needed for Fabricating Silicon Carbide-based Injector Components" in "Advanced Processing and Manufacturing Technologies for Structural and Multifunctional Materials II" edited by T. Ohji and M. Singh, CESP, Vol. 29, Issue 9, pp. 1-14, Wiley, NY and American Ceramic Society (2009).
[11]M.C. Halbig and M. Singh, "Diffusion Bonding of Silicon Carbide for the Fabrication of Complex Shaped Ceramic Components," Ceramic Integration and Joining Technologies: From Macro- to Nanoscale, Eds. M. Singh, T. Ohji, R. Asthana and S. Mathur, John Wiley & Sons, 2011.
[12]J. Ayanne, L. Beaunier, J. Boumendil, G. Ehret and D. Laub, Sample Preparation Handbook for Transmission Electron Microscopy: Techniques, 1st Edition, P. 135, Springer, N.Y., 2010.
[13]J. H. Schneibel and C. J. Rawn, "Thermal Expansion Anisotropy of Ternary Silicides Based on Ti$_5$Si$_3$," Acta Materialia, Vol. 52 (2004) 3843-3848.
[14]L. Zhang and J. Wu, "Thermal Expansion and Elastic Moduli of the Silicide Based Intermetallic Alloys Ti$_5$Si$_3$(X) and Nb$_5$Si$_3$," Scripta Materiallia, Vol. 38, Issue 2 (1998) 307-313.

JOINING OF ALUMINA BY USING OF POLYMER BLEND AND ALUMINUM

Ken'ichiro Kita, Naoki Kondo, and Hideki Kita
Advanced Manufactureing Research Institute,
National Institute of Advanced Industrial Science and Technology (AIST)
Nagoya, Aichi, Japan

Yasuhisa Izutsu
Stereo Fabric Research Association
Nagoya, Aichi, Japan

ABSTRACT

Alumina joining using Al foil and a polymer blend containing polycarbosilane (PCS) and polymethylphenylsiloxane (PMPhS) was carried out, and the average flexural strength was approximately 239 MPa. The joining layer of this sample contained metal Si because of a reduction process between the Al and polysiloxane. On the other hand, alumina joining using Al foil and a polymer blend containing PCS and polymethylsilsesquioxane (PMSQ) was also carried out, and the average flexural strength was approximately 176 MPa. The joining layer of this sample contained metal Al and needle-shaped metal Si that was presumably derived from partial reduction of Si oxides included in PMSQ.

INTRODUCTION

The increasing environmental impact is one of the most serious problems worldwide, and it is desirable to find ways to decrease it.   We maintain that replacing certain materials used in high-temperature production processes that cause the environmental impact (e.g., rare metals), with ceramics is a good solution to this problem.[1] However, such components are usually complex and large-scaled. Therefore, it is very difficult to make ceramics of such components because a large furnace is required for their fabrication, and they are expensive to prepare. To facilitate their use, we investigated a method for making numerous precision hollow units and then assembling them by joining. In this method, a novel joining method is required.

We gave attention to ceramic joining using a ceramic precursor polymer such as polycarbosilane (PCS) because of the superior characteristics of this method, including its low cost compared with grinding, low impurities compared with glass joining, and applicability to complex and small ceramic units. There have been many reports on joining using such a polymer.[2-5] However, the drawback of this method is that the tensile strength of a sample joined by such a polymer is not very high, as compared to any other joining method such as metallization, glass joining, or solid-state welding. Therefore, we attempted to perform joining using PCS and aluminum foil in anticipation of a direct reaction between SiO_2 and metal aluminum. PCS could be made into SiO_2 by oxidation curing and heating, and a direct reaction made it possible to form aluminum-silicon oxide by controlling the heating temperature, which increased the tensile strength of the joined sample.[6]

Based on this result, we attempted to use polysiloxane instead of PCS, because polysiloxane includes a large quantity of oxygen, which could save the processes of oxidation curing and heating during the joining by this method. Moreover, polysiloxane is less expensive than PCS. Polysiloxane and alumina cannot be joined directly. Therefore, we prepared a polymer blend containing PCS and polysiloxane, because PCS can react to alumina and no exfoliation occurs between PCS and alumina.[7] In this study, we investigated whether the use of such a polymer blend was effective and what kind of effect it had.

EXPERIMENTAL PROCEDURE

Figure 1 shows the outline of the experimental procedure used in this study. Bulk alumina (with a purity of more than 99.9 %) supplied by Mitsui Kinzoku Corporation was cut into alumina pieces (length: 20 mm, width: 30 mm, height: 20 mm). A surface with an area of 600 mm^2 was abraded using a grinder, and the following investiga   tions were carried out on the abraded surface.

In this experiment, two types of polysiloxanes were prepared, and their chemical structures are shown in Figure 2. This first is polymethylphenylsiloxane (PMPhS), which contains the methyl and phenyl groups in the side chains. The other is polymethylsilsesquioxane (PMSQ), which is a cage-shaped polysiloxane polymer (a type of silicon resin) and other groups connected between the cage-shaped polysiloxanes. Commercial PMPhS (KF-54, Shin-Etsu Chemicals Co. Ltd., Japan) was blended with PCS (NIPUSI-Type A, Nippon Carbon, Japan) at a PMPhS to PCS blend ratio of 30 mass%. The blended polymer was then dissolved into dehydrated toluene. Commercial PMSQ (YR-3370, Momentive Performance Materials Japan Inc., Japan) was blended with PCS (NIPUSI-Type A, Nippon Carbon, Japan). The PMSQ to PCS blend ratio and the solvent for the blended polymer were the same as the case using PMPhS.

The abraded lower parts of the alumina pieces were dipped in the toluene solution with the above polymer blends a few times and then dried. Thereafter, the alumina pieces dipped into the

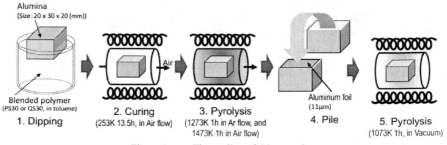

Figure 1.   The outline of this experiment

polymer blend containing PCS and PMPhS were referred to as PS30, and the alumina pieces dipped into the polymer blend containing PCS and PMSQ were referred to as QS30. After this process, these pieces were cured at 253 K for 13.5 h under air flow, pyrolyzed at 1273 K for 1 h

Polymethylphenylsiloxane (PMPhS)   Polymethylsilsesquioxane (PMSQ)

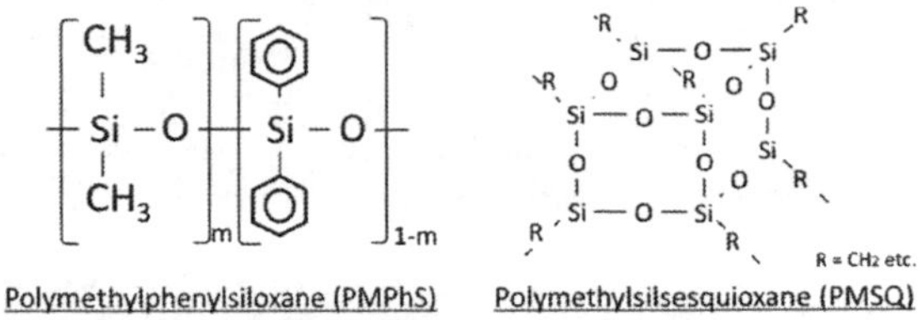

Figure 2.   Chemical structures of PMPhS and PMSQ

under air flow, and heated at 1473 K for 2 h. The curing and pyrolysis induced ceramization of the polymer without mass reduction and prevented exfoliation between the alumina and polymer.[8] In addition, excess carbon was eliminated from the polymer by heating.

After heating, two of the pieces were selected for experimentation. An aluminum foil with a thickness of approximately 11 μm was placed between the abraded surfaces of the pieces, and the pieces were heated at 1073 K for 2 h in vacuum.

These samples were then investigated using scanning electronic microscopy (SEM; JEM-5600, JEOL, Japan), energy dispersive X-ray spectroscopy (EDS; JEM-5600, JEOL, Japan), and X-ray diffraction (XRD; RINT2500, Hitachi Ltd., Japan) for observing the joining area of the samples. In addition, a four-point flexural test was performed for estimating the strength of the samples following JIS R1601. The upper and lower spans were 10 and 30 mm long, respectively.

RESULTS AND DISCUSSION

To clarify the compositions in the joining area, XRD spectra of the joining areas of PS30 and QS30 were obtained. These spectra are shown in Figure 3. The spectrum of PS30 showed peaks at 38.3° and 44.5°, suggesting the existence of metal Al, and the peaks at 28.4° and 46.2° suggested the existence of metal Si. There were wide and continuous peaks for θ = 10.0°–14.0° and around 20.0°. Such peaks prove the existence of an amorphous compound.[9] It is well-known that polymer pyrolysis forms tiny and pure crystals which are covered by an amorphous material.[10] Therefore, the existence of this peak is reasonable. As for the spectrum of QS30, the remarkable peaks were almost the same as those of PS30, and the joining area appeared to include metal Al, metal Si, and amorphous.

Figure 4 shows SEM images and EDS mappings of the joining area of PS30 and QS30. In the case of PS30, the SEM image shows that the color of the joining area was lighter than the alumina, and the width was approximately 30 μm. The EDS mappings show that the joining area was almost occupied by a strong Si spectrum, and the other spectra were barely measurable. Therefore, it was considered that this joining area mainly consisted of metal Si.[11] As for QS30, the color of the joining area was also light compared to the color of alumina, and the width was also approximately 30 μm. EDS mappings show that the Al spectrum mainly occupied this area, while the Si spectrum in this area was derived from numerous needle-shaped particles that were diffused there. In addition, the Al spectrum was not observed in the area where the Si particles existed. This revealed that this joining area was mainly occupied by metal Al, and needle-shaped metal Si was diffused in the metal Al. The results in Figure 3 show that both joining areas included metal Al and metal Si. Therefore, the above consideration appeared to be reasonable.

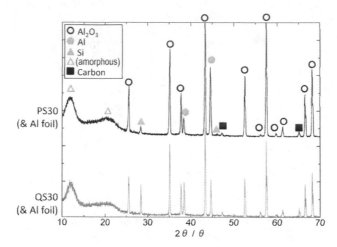

Figure 3.   XRD spectra of the joining area of PS30 and QS30

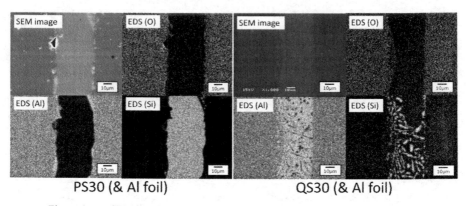

Figure 4.   SEM image and EDS mappings of the cross section of the joining area

The strength of these joining areas was measured using a four-point bending test. The results are shown in Figure 5, and the related detail data are summarized in Table I. In addition, Figure 6 shows the Weibull distributions of the results of the four-point bending tests of these samples.

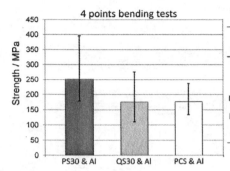

Table I. Detail result of the 4 points bending tests

Sample (& Al foil)	PS30	QS30	PCS (Comparison)
Sample Number	7	7	12
Average (MPa)	239	176	177
Maximum Strength (MPa)	396	276	237
Minimum Strength (MPa)	160	111	134
Weibull distribution	2.87	2.98	5.49

Figure 5.   The result of the 4 point bending tests

The case of joining by using PCS and aluminum foil only is provided in this graph as a comparison. In the case of PS30, the average strength was approximately 239 MPa, and the maximum strength was approximately 396 MPa. The maximum value was the same as that of alumina because the fracture in the strength test for this sample occurred in the alumina rather than the joining area. In the case of QS30, the average strength was approximately 176 MPa, which was almost the same as that of PCS. The maximum strength of QS30 was higher than that of PCS while the minimum was lower. Figure 6 shows that the Weibull modulus of QS30 was lower than that of PCS. It reveals that the strength of QS30 was more uneven than that of PCS, although the maximum strength was higher than that of PCS.

We considered why these differences occurred between PS30 and QS30 despite including a large quantity of Si-O-Si bonds, focusing on the behaviors of PMPhS and PMSQ during heating. PMPhS is one of the straight silicone fluids. Therefore, it can be decomposed and reconstructed into low molecular siloxanes at more than 523 K, and the siloxanes can be dissolved into any other polymer.[12] In this case, PMPhS appeared to be dissolved into PCS because PCS can melt at more than 523 K.[13] Moreover, a paper reported that there was no remarkable unevenness in reduction of Si oxides in SiC fibers made from PCS, including low molecular siloxanes derived from straight silicone fluid.[14] Therefore, it is considered that low molecular siloxanes could be dissolved into PCS uniformly; the uniform reduction of Si oxides could cause a uniform reaction with metal Al, and the single metal Si joined to the alumina. On the other hand, PMSQ can melt at 400 K. However, it becomes hard again after 40 min at this temperature and cannot be melted into PCS uniformly.[15] That is, Si oxides in PCS including PMSQ also are partially reduced, resulting in the needle shape of the metal Si, and hence a large quantity of metal Al unreacted with Si oxides remains in the joining area.

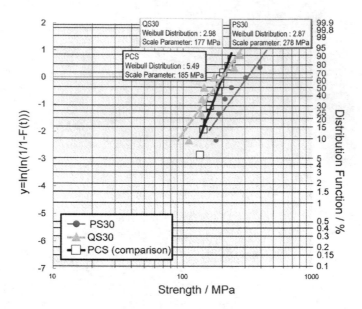

Figure 6.   Weibull distribution of the 4 point bending tests of the samples

CONCLUSION

In the case of alumina joining using Al foil and a polymer blend containing PCS and PMPhS, the average tensile strength of the joined sample was approximately 239 MPa, and the sample was mainly joined by metal Si. It was considered that the metal Si joining was brought about by the uniform reduction of Si oxides derived from the thermal decomposition and dissolution of PMPhS. As for the case of alumina joining using Al foil and a polymer blend containing PCS and PMSQ, the average tensile strength of the joined sample was approximately 176 MPa, and the sample was mainly joined by metal Al and needle-shaped metal Si. It was thought that the partial resuction of Si oxides derived from the thermal re-hardening and incomplete dissolution of PMSQ formed the complex joining area.

ACKNOWLEDGEMENT

This research was supported by METI and NEDO, Japan, as part of the Project for the Development of Innovative Ceramics Manufacturing Technologies for Energy Saving.

REFERENCE

[1] H. Kita, H. Hyuga, N. Kondo, and T. Ohji, Exergy Analysis on the Life Cycle of Ceramic Parts, *Key Eng. Mater.*, **403**, 261-264 (2009).

[2] S. Yajima, K. Okamura, T. Shishido, Y. Hasegawa, and T. Matsuzawa, Joining of SiC to SiC using polybolosiloxane, *Am. Ceram. Soc. Bull.*, **60**,253 (1981).

[3] E. Anderson, S. Ijadi-Maghsoodi, O. Ünai, M. Nostrati, and W. E. Bustamante, Development of a compound for low temperature joining of SiC ceramics and CFCC composites, *Ceram. Trans.*, **77**, 25-40 (1997).

[4] P. Colombo, V. Sglavo, E. Pippel, and A. Donato, Joining of reaction-bonded silicon carbide using a preceramic polymer *J. Mater. Sci.*, **33**, 2405-2412 (1998).

[5] P. Colombo, Joining Ceramics using Preceramic Polymers, in A. Bellosi, T. Kosmač, and A. P. Tomsia (eds.), *Interfaccial Science in Ceramic Joining*, Klwer Academic Publishers, 405-413 (1998).

[6] S. Maitra, R. Shibayan, and A. K. Bandyopadhyay, Synthesis of mullite from calcined alumina, silica and aluminium powder, *Ind. Ceram.*, **24**, 39-42 (2004)

[7] K. Kita, N. Kondo, Y. Izutsu, and H. Kita, Investigation of the properties of SiC membrane on alumina by using polycarbosilane, *Mater. Lett.*, accepted.

[8] Y. Hasegawa, M. Iimura, and S. Yajima, Synthesis of continuous silicon carbide fibre Part2 Conversion of polycarbosilane fibre into silicon carbide fibres, *J. Mater. Sci.*, **15**, 720-728 (1980).

[9] Y. Hasegawa, Synthesis of continuous silicon carbide fibre Part6 Pyrolysis process of cured polycarbosilane fibre and structure of SiC fibre, *J. Mater. Sci.*, **24**, 1177-1190 (1989).

[10] B. B. Straumal, S.G. Protasova, A. A. Mazilkin, B. Baretzky, A. A. Myatiev, P. B. Straumal, Th. Tietze, G. Schütz and E. Goering, Amorphous interlayers between crystalline grains in ferromagnetic ZnO films *Mater. Lett.*, **71**, 21-24 (2012)

[11] T. Okutani, Utilization of Silica in Rice Hulls as Raw Materials for Silicon Semiconductors, *J. Metal. Mater. Mineral*, **19**, 51-59 (2009)

[12] K. Kita, M. Narisawa, H. Mabuchi, M. Itoh, M. Sugimoto, and M. Yoshikawa, Formation of continuous pore structures in Si-C-O fibers by adjusting melt spinning condition of polycarbosilane - polysiloxane polymer blend, *J. Am. Ceram. Soc.*, **92**, 1192-7 (2009).

[13] A. Idesaki, M. Narisawa, K. Okamura, M. Sugimoto, S. Tanaka, Y. Morita, T. Seguchi, and M. Itoh, Fine SiC fiber synthesized from organosilicon polymers: relationship between spinning temperature and melt viscosity of precursor polymers, *J. Mater. Sci.*, **36**, 5565-69 (2001).

[14] K. Kita, M. Narisawa, H. Mabuchi, A. Nakahira, M. Sugimoto, and M. Yoshikawa, Synthesis and properties of ceramic fibers from polycarbosilane/polymethylphenylsiloxane polymer blends, *J. Mater. Sci.*, **45**, 3397-3404 (2010).

[15] M. Narisawa, R. Sumimoto, K. Kita, H. Kado, H. Mabuchi, and Y. –W. Kim, Melt Spinning and Metal Chloride Vapor Curing Process on Polymethylsilsesquioxane as Si-O-C Fiber Precursor, *J. Appl. Polym. Sci.*, **114**, 2600-2607 (2009).

DIFFUSION BONDING OF RIGID ALUMINA PIECES USING POROUS ALUMINA INTERLAYERS

Hiroyuki Miyazaki, Mikinori Hotta and Hideki Kita
National Institute of Advanced Industrial Science and Technology (AIST)
Anagahora 2266-98, Shimo-shidami, Moriyama-ku, Nagoya 463-8560, Japan

Yasuhisa Izutsu
Stereo Fabric Research Association
Anagahora 2268-1, Shimo-shidami, Moriyama-ku, Nagoya 463-0003, Japan

ABSTRACT
A pure alumina slurry with dispersant was sandwiched by an alumina plate couple and was dried in the gap of faying surfaces, followed by pressureless sintering at 1650 °C. An average of flexural strength of more than 260 MPa was attained at room temperature for the joint prepared from a slurry with solid content of 42.2 vol%, although the porosity of the interlayer was ~20%. Furthermore, a high temperature strength of more than 150 MPa was achieved at 1200°C in air for this sample. It was found that diffusion bonds of alumina could be made without either joining pressure or perfectly flat surfaces by controlling the interlayer microstructure and nature of the flaws.

INTRODUCTION

Direct ceramic-ceramic joining by solid diffusion is one of the attractive methods since this technique has the ability not to adversely affect the high temperature properties of the joint. However, high pressures during the heat treatments and/or highly flat faying surfaces are necessary.[1] In order to reduce the temperature and pressure needed for the ceramic-ceramic joining, submicron and nanocrystalline grained ceramics interlayers were used.[2-6] From the point of practical application to large and complex components, however, those techniques are still unsatisfactory since bonding without external pressure is ideal and the applicability to non-flat faying surfaces is of importance as well. An attempt to fabricate an interlayer using slurry coating of nano powder was carried out by Cross and Mayo to achieve the latter goal.[2] They regarded the resulting joints as unsuccessful due to the substantial shrinkage void formation in the joint upon sintering. Their conclusion seems fair since it is generally believed that a porous interlayer deteriorates the mechanical properties of the joint. However, the flexural strengths of porous alumina reported by Nanjangud et al. exceeded 100 MPa when the relative density was 70% and above.[7] It is rational to suppose that the porous alumina interlayer possesses enough joining strength if large flaws are avoided.

In this study, diffusion bonding of commercial alumina was examined with a pure alumina interlayer at the joint. The interlayers were applied in the form of a slurry where the solid loading was 42.2 or 46.3 vol%. After drying the slurry in the gap of faying surfaces, the green joints was heat treated at a temperature of 1650°C. Joint strength was measured at both room temperature and 1200°C and discussed in conjunction with the microstructures of the joints.

EXPERIMENTAL PROCEDURE

Commercially available alumina ceramics (Hi-Cera HA, Mitsui Mining & Smelting Co., Ltd.) were used as the specimen to be joined. The specimen possessed a purity of 99.6 mass%, a density of 3.91 g/cm^3 and a room temperature bending strength of 380 MPa. Alumina plates with a dimension of 20 x 16 x 5 mm were prepared from the alumina ceramics, and the larger surfaces were ground with a 200-grit diamond wheel before joining tests. Low soda alumina powder (0.6 μm, AL-

160SG-4, Showa Denko K. K.) with less than 0.2 mass% total impurity content and distilled water were the starting materials for the slurry to form the joining interlayer.   A dispersant (Serna D305, Chukyo Yushi Co. LTD.) was added to the distilled water at 5.6 mass % before mixing the alumina powder. Both dilute slurry (hereafter DS) and concentrated slurry (CS) were prepared, which solid content were 42.2 vol% and 46.3 vol%, respectively.   The mixed slurry was evacuated for 2 min with a vacuum pump to eliminate bubbles and then a few drops of the slurry were spread over the 20 x 16 mm surface of the alumina plate.   The other alumina piece was set on top of the alumina slurry so that the gap between faying surfaces was ~90 μm, followed by drying overnight at room temperature. The joint with the gap of ~60 μm was also fabricated for the DS sample to study the effect of the width of the gap (DS2/3).   The dried assemblies were heated in air at 1650 °C for 2 h with a negligible pressure of 0.03 MPa by placing a suitable weight on top of the substrate.

The densities of the green interlayer after drying were obtained from their thickness, area and weight, all of which were measured after disassemble of the green interlayer.   Relative densities of the green interlayer were calculated by using a theoretical density of 3.99 g/cm^3 for alumina. Apparent relative densities of the interlayer after sintering were estimated from both the relative density of the dried green interlayer and the shrinkage of the gap between faying surfaces due to sintering.

The joined samples were cut perpendicularly to the interlayer and polished with a 0.5- μm diamond slurry, followed by thermal etching before microstructural observation by scanning electron microscopy (SEM).   Specimens with dimension of 2 x 3 x 10 mm were machined for strength measurements.   These specimens have joints at the center of bend bars with the joints perpendicular to the longitudinal direction.   Three-point bending strength was measured with a span of 8 mm and a crosshead speed of 0.5 mm/min.   In addition to the test at room temperature, high temperature bending test at 1200°C in air was performed for DS sample since it exhibited good flexural strengths at room temperature.   In this case, alumina blocks with a dimension of 20 x 40 x 13 mm were cut from the alumina ceramics.   Both surfaces of the 40 x 13 mm planes of the plates were ground with a 200-grit grinding wheel before joining tests.   The joining process was the same as that for the small specimens. Four standard-size specimens (3 x 4 x 40 mm) were prepared from the joint sample.[8] Four-point bending strength was measured at 1200°C in air in accordance with JIS R 1601 using an inner and outer span of 10 and 30 mm and a crosshead speed of 0.5 mm/min.

RESULTS AND DISCUSSION

Microstructure of the alumina joints

Apparent relative density of the interlayer after sintering is shown in Table I, which indicated that the interlayers of the three joints possessed porosities more than 20%.   Figure 1 shows the microstructure of the alumina joints for DS, DS2/3 and CS samples prepared from pure alumina slurry with solid contents of 42.2, 42.2 and 46.3 vol%, respectively.   The microstructures of the three samples resembled each other except the existence of lateral interfacial cracks in the last sample.   The size of interfacial cracks in the CS sample often exceeded more than 300 μm.   The mechanism of the formation of interfacial cracks in the CS sample was not clear at this point.   It is likely that the interfacial crack developed during the sintering since the dried green interlayer bonded so firmly to the faying surfaces that the assemblies could not be separated easily by hand.   Further study is needed to clarify the origin of these large interfacial cracks.

The common feature of the three samples was that the interlayers consisted of three characteristic parts: a dense area, a porous area and large voids.   Figure 2 is the enlarged pictures of these three constituent parts in the DS sample.   Many small pores with size of around 2 μm existed in the left side of figure 2(a) to form porous area, while few pores resided in the right side of the figure.

Interestingly, these dense areas often located next to the large voids. A typical example of large voids is presented in figure 2(b). The heights of most voids were the same as the thickness of the interlayer. The widths of ~60 coarse voids were measured from several SEM pictures for each sample and their distributions are shown in figure 3. The width of voids in both DS and CS samples ranged from 10 μm to ~100 μm and their average sizes were ~41 μm. By contrast, the range of width distribution of voids for the DS2/3 samples shifted to smaller side than that of the DS sample although some large voids with width of ~100 μm were observed occasionally.

Table I. Apparent relative density and bending strength of the alumina joints prepared from pure alumina slurry

Material code	Volume fraction of alumina in the slurry (%)	Apparent relative density of interlayer (%)	Bending strength at room temperature (MPa)	Bending strength at 1200 °C (MPa)
DS2/3	42.2	78	269 ± 33	-
DS	42.2	80	265 ± 40	157 ± 53
CS	46.3	75	119 ± 12	-

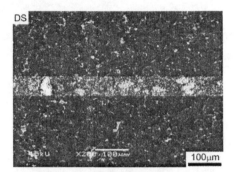

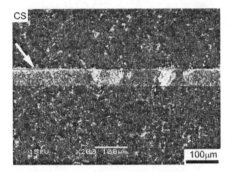

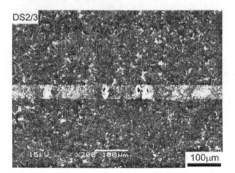

Figure 1. SEM micrographs of the alumina joints prepared from pure alumina slurry with solid content of 42.2 vol % (DS and DS2/3) and 46.3 vol% (CS). A dense area, a porous area and large voids were observed in the interlayer. The white arrow indicates an interfacial crack between the interlayer and the original alumina plate, which existed only in the CS sample.

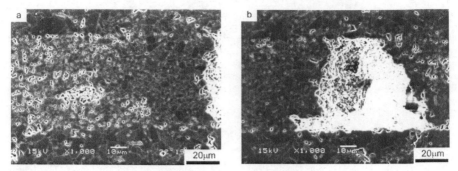

Figure 2. SEM micrographs of the characteristic parts in the alumina joints observed at a high magnification of 1k. The interlayer was prepared from alumina slurry with solid content of 42.2 vol % (DS). (a) a porous area (left side) and a dense area (right side), (b) a large void.

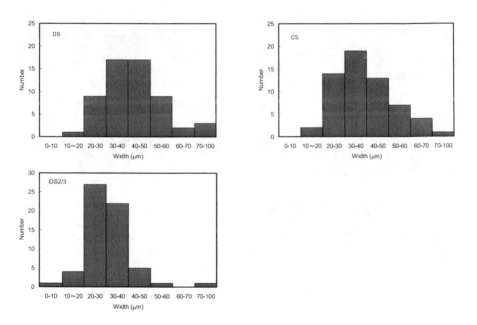

Figure 3. Width distributions of voids in the alumina joints prepared from pure alumina slurry with solid contents of 42.2 vol% (DS and DS2/3) and 46.3 vol% (CS). The total number of voids counted was ~60 for each sample.

Bending strength of the alumina joints

Table I presents three-point bending strength of the alumina joints. The number of the specimen tested at room temperature was four for the DS2/3 sample and three for the DS and CS samples. The DS2/3 and DS samples exhibited an average strength of more than 260 MPa, whereas that of the CS sample decreased significantly to 119 MPa. It is well known that the mechanical properties of the porous ceramics depend on their porosities. It is likely that the similar bending strength for the DS2/3 and DS samples resulted from almost the same porosity of the interlayer. The inferior joining strength of the CS sample was attributable to the large interfacial cracks in it since the sizes of large voids in the CS sample were almost the same as those in the DS sample and hardly exceeded 100 μm. It has been reported that four-point bending strength of partially sintered porous alumina with the relative density of 0.73 was about 110 MPa and it increased up to about 260 MPa as the relative density approached to 0.92.[7] Thus, the bending strength of the DS2/3 and DS sample seems to be reasonable when compared with that of porous alumina with similar porosity. In the case of joining with glasses, reduction of the interlayer thickness results in stronger joints [9,10] The similar bending strengths for the DS2/3 and DS samples regardless of the interlayer thickness can be explained by the lack of thermal expansion mismatch between the adhesive and adherend which is the origin of the dependence of strength on the thickness of glass interlayer.

The literatures demonstrate that the bending strengths of most alumina joints with glass interlayer range from 150 to 260 MPa.[11-15] Therefore, the flexural strengths of the DS and DS2/3 samples were comparable to the best results previously reported for alumina joints with glasses. The bending strength at 1200°C for the DS sample was more than 150 MPa, which demonstrated its advantage over the alumina joints with glass interfaces since most of the softening temperatures of those glasses were below 1100°C. [11,12,14]

There have been few studies on the diffusion bonding of oxide ceramics using the oxide slurry. Cross and Mayo employed zirconia slurry for diffusion bonding of zirconia plates and found that large shrinkage voids were formed in the interlayer even when a compressive pressure of 10 MPa was applied during the sintering.[2] They considered the results unsuccessful and no joining strength was measured for their samples since they believed that such large voids in the interlayer inevitably reduced the strength of the joint. It is probable that mechanical strength of their joints was poor as they expected since they abandoned to optimize the process condition. By contrast, the slurry method was improved in the following two ways in our studies since we inferred that a porous interlayer could sustain good adhesion strength if large strength-limiting flows were eliminated. The first point is the optimization of the slurry concentration and the second one is the procedure, that is, the slurry was dried in the gap of faying surfaces to connect the green interlayer tightly to the alumina plates, while in the previous work both plates were coated with a slurry and assembled after drying. It was supposed that these two points were the key improvements for realizing the high bonding strength of the DS sample.

Our results of the DS and DS2/3 samples indicated that the bending strength of the joints was not so susceptible to the thickness of the interlayer, which would be beneficial to the joining of large components where smooth and flat faying surface can not be prepared easily. In order to clarify the tolerable amount of the gap between the adherend for this slurry method, the effect of the interlayer thickness will be investigated widely in near future.

CONCLUSION

Generally, large voids at the joints are believed to deteriorate the bonding strength, so that both high pressure at elevated temperature and highly smooth faying surfaces are employed for the diffusion bonding of ceramics to eliminate the voids in the interlayer. This study, however, revealed that a porous interlayer possessed sufficient adhesion strength provided that macro defects, which seriously limit the strength, were eliminated. The unique point of our technique is that a pure alumina slurry was used for the source of the porous interlayer. The slurry was sandwiched by an alumina plate couple and was pressurelessly sintered at 1650°C for 2 h after drying between two adjoining dense alumina plates. It was found that large interfacial flaws could be removed from the porous interlayer when the concentration of the slurry was properly controlled. With this finding, diffusion bonding of pure alumina with high flexural strengths of ~ 260 MPa at room temperature and 157 MPa at 1200°C could be successfully achieved without external pressure. The unsusceptibility of the joint strength to the interlayer thickness implied that the method might be applicable to large components with uneven surfaces.

ACKNOWLEDGEMENTS

This research was supported by METI and NEDO, Japan, as part of the Project for the Development of Innovative Ceramics Manufacturing Technologies for Energy Saving.

REFERENCES

[1] J. A. Fernie, R. A. L. Drew and K. M. Knowles, Joining of engineering ceramics, *Int. Mater. Rev.* **54**, 283-331 (2009).

[2] T. H. Cross, M. J. Mayo, Ceramic-ceramic diffusion bonding using nanocrystalline interlayers, *Nanostruct. Mater.*, **3**, 163-68 (1993).

[3] H. Ferkel, W. Riehemann, Bonding of alumina ceramics with nonoscaled alumina powders, *Nanostruct. Mater.*, **7**, 835-45 (1996).

[4] R. J. Hellmig, J. -F. Castagnet and H. Ferkel, Stability of alumina ceramics bonded with nonoscaled alumina powders, *Nanostruct. Mater.*, **12**, 1041-44 (1999).

[5] R. J. Hellmig, H. Ferkel, Using alumina nonopowders as cement in bonding of alumina ceramics, *Phys. Stat. Sol.(a)*, **175**, 549-53 (1999).

[6] B. Cina, I. Eldror, Bonding of stabilized zirconia (Y-TZP) by means of nano Y-TZP particles, *Mater. Sci. Eng.*, **A301**, 187-95 (2001).

[7] S. C. Nanjangud, R. Brezny and D. J. Green, Strength and Young's modulus behavior of a partially sintered porous alumina, *J. Am. Ceram. Soc.*, **78**, 266-68 (1995).

[8] Testing methods for flexural strength (modulus of rupture) of fine ceramics, Japanese Industrial Standard, JIS R 1601, 1995.

[9] S. M. Johnson, D. J. Rowcliffe, Mechanical properties of joined silicon nitride, *J. Am. Ceram. Soc.*, **68**, 468-72 (1985).

[10] W. A. Zdaniewski, J. C. Conway, JR., and H. P. Kirchner, Effect of joint thickness and residual stresses on the properties of ceramic adhesive joints: II, Experimental results, *J. Am. Ceram. Soc.*, **70**, 110-18 (1987).

[11] S. Fujitsu, S. Ono, H. Nomura, M. Komatsu, K. Yamagiwa, E. Saiz and A. P. Tomsia, Joining of single-crystal sapphire to alumina using silicate glasses, *J. Ceram. Soc. Jpn.*, **111**, 448-51 (2003).

[12] L. Esposito, A. Bellosi, Ceramic oxide bonds using calcium aluminosilicate glasses, *J. Mater. Sci.*, **40**, 2493-98 (2005).

[13] B. G. Ahn, Y. Shiraishi, SiO_2-CaO-Al_2O_3 glass solder for joining of Al_2O_3 to Al_2O_3, *High Temp. Mater. Process.*, **17**, 209-16 (1998).

[14] J. T. Klomp, Th. P. J. Botden, Sealing pure alumina ceramics to metals, *J. Am. Ceram. Soc. Bull.*, **49**, 204-11 (1970).

[15] J. T. Knapp, J. D. Cawley, Silicate brazing of alumina ceramics using calcium aluminosilicate interlayers, in: P. Kumar, V.A. Greenhut (Eds.), Metal-Ceramic Joining, The Minerals, Metals & Materials Society, 1991, pp. 181-201.

INFLUENCE OF JOINING PRESSURE AND SURFACE ROUGHNESS ON FLEXURAL STRENGTH OF JOINED BORON CARBIDE CERAMICS

Kiyoto Sekine[a], Takeshi Kumazawa[b], Wu-Bian Tian[a], Hideki Hyuga[c] and Hideki Kita[c]
[a]Stereo Fabric Research Association, Nagoya, Japan
[b]Technical Research Laboratory, Mino Ceramic Co., Ltd. Handa, Aichi, Japan
[c]National Institute of Advanced Industrial Science and Technology (AIST), Nagoya, Japan

ABSTRACT
    The influence of the joining pressure and surface roughness on the flexural strength of boron carbide (B_4C) joined by an Al sheet was investigated. The joining surface of B_4C was ground or polished to a surface roughness ranging from 0.1 to 6.1 μm. B_4C was joined by Al under a pressure of 0.5–12 kPa or stabilized with a carbon jig at 1000 °C for 2 h in vacuum (10^{-2}–10^{-4} Pa). The fracture surfaces were observed by an optical microscope and the microstructures in the B_4C joining layer were observed by scanning electron microscopy. The flexural strength of B_4C joined at 12 kPa and stabilized with a carbon jig was similar to that of the B_4C base material. However, the flexural strength of B_4C joined at less than 6 kPa was lower than that of the B_4C base material because of the presence of voids in the joining interlayer. When the surface roughness was 6.1 μm, the flexural strength was slightly lower than that of the B_4C base material. However, when the joining surface roughness ranged from 0.1 to 1.7 μm, the flexural strength of joined B_4C was similar to that of the B_4C base material.
Keywords: boron carbide, joining, aluminum, pressure, surface roughness

INTRODUCTION
    Boron carbide (B_4C) ceramics are an important type of non-oxide ceramic with high hardness, high elastic modulus, and low density.[1] On account of these good physical properties, they have been widely used in the manufacture of semiconductor production equipment, wear-resistant components, super-hard tools, and armor.[2] Nowadays, large-sized B_4C ceramics with complicated shapes are required for such applications. This is because of changing requirements; for example, the size of semiconductor manufacturing equipment has increased to enhance throughput, and in the near future, Si wafer sizes are expected to increase from the current 12 in to 16 in. The joining of B_4C ceramics is considered one of the promising and effective approaches for producing B_4C components for such applications.
    Cockeram[3] previously investigated the joining of B_4C ceramics. In this study, B_4C was joined by refractory metal foils such as those of Ti and Mo for high-temperature applications. He tested the joining of B_4C at 1500 °C for 10 h under a pressure of 18.6 MPa by using various foils, and obtained microstructural observations indicating that Mo foil was the most suitable bonding material for B_4C joining. In contrast, for low-temperature applications such as the manufacture of Si wafers, a low-melting-point metal phase can instead be used to realize lower joining temperatures. However, no studies have focused on the joining of B_4C at low temperatures. It is well known that Al has a low melting point of 660 °C and is an effective sintering additive for the preparation of B_4C.[4-6] Kumazawa et al.[4] reported that B_4C compacts were pressureless sintered with Al gas and Si compound gas derived from SiC. Miyazaki et al.[5] studied the microstructure of B_4C pressureless sintered in Ar atmosphere containing Al gas and Si compound gas. They concluded that Al was an effective sintering additive for the preparation of B_4C ceramics. In addition, the reaction between B_4C and Al has been studied extensively to determine the wetting properties and from the viewpoint of applications to the manufacture of composites.[7-14] Owing to these studies, we focused on Al as the joining material for B_4C.

Generally, the important parameters for joining include temperature, joining time, pressure, and roughness of joining interface. Suganuma et al.[15] studied the relationship between joining pressure and joining strength of Si_3N_4 by using Al and the relationship between surface roughness and joining strength. They reported that the Weibull slope of the joints bonded at 0 MPa was 2.8, whereas that of the joints bonded at 0.05 MPa was 14.8. In addition, Nagano et al.[16] reported the relationship between surface roughness and bonding strength of tetragonal ZrO_2 including 3 mol% Y_2O_3. They reported that bonding strength decreased as the bonding surface roughness increased. However, neither the relationship between joining strength and pressure, nor that between strength and surface roughness, has been clarified for B_4C joining.

In this study, tests of B_4C joining by Al and microstructural observations of the joining interlayer were carried out to clarify the influences of the joining pressure and surface roughness on the flexural strength of B_4C ceramics joined by using Al.

EXPERIMENTAL

B_4C ceramic plates with a density of 2.40 g/cm^3 (95.2% of the theoretical density) and a room-temperature bending strength of 398 ± 74 MPa were purchased (Mino Ceramic, Nagoya, Japan) and used as base materials (dimensions: 20 mm × 25 mm × 5 mm) for the joining tests. The surface of the B_4C plate (25 mm × 5 mm) was ground with a diamond wheel or polished for a mirror surface, and the average roughness, Ry, of the joining surface was measured by a surface roughness meter (Surfpack-Pro; Mitutoyo, Kanagawa, Japan). The joining surface roughness of the mirror-polished surface was 0.1 μm, and that of the ground surface was 1.0–6.1 μm. An Al sheet (thickness: 10 μm; purity: >99.0%; Nilaco, Tokyo, Japan) was sandwiched between the ground B_4C ceramic plates. Weights were loaded on the B_4C–Al–B_4C sandwich. The joining pressure ranged from 0.5 (no load) to 12 kPa. Alternatively, the B_4C–Al–B_4C sandwich was stabilized by using a jig made of carbon (ET-10; Ibiden, Gifu, Japan). Joining was conducted in a furnace at 1000 °C for a soaking time of 2 h in vacuum (10^{-2}–10^{-4} Pa).

The joined samples were cut into pieces with dimensions of 3 mm × 4 mm × 40 mm, which were ground with a diamond wheel until the roughness Ra was less than 0.2 μm. Their flexural strength was measured by a four-point bending test with an inner and outer span of 10 mm and 30 mm, respectively, at a displacement rate of 0.5 mm s^{-1} on a universal testing machine (Sintech 10/GL; MTS Systems, USA). Three samples were used for each joining condition. Six samples were tested for the B_4C base material.

The fracture surface was observed by an optical microscope. The joining interfaces were polished by using a cross-section polisher (SM-09010; JEOL, Tokyo, Japan), and then observed by field-emission scanning electron microscopy (FE-SEM) at 7.5 kV (JSM-7000F; JEOL, Tokyo, Japan).

RESULTS AND DISCUSSION

Dependence of flexural strength on joining pressure

Figure 1 shows the dependence of the flexural strength of joined B_4C on the joining pressure. The flexural strength of the B_4C base material was about 400 MPa. The flexural strength of B_4C joined at 0.5 kPa (no load) and 6 kPa ranged from 270 to 340 MPa. In addition, the flexural strength of B_4C joined at 12 kPa and stabilized with a carbon jig ranged from 380 to 420 MPa. Thus, the strength of B_4C joined at 12 kPa and stabilized with a carbon jig was close to that of the B_4C base material, whereas the strength of B_4C joined at 0.5 kPa or 6 kPa was lower than the latter.

Figure 2 shows typical fracture surfaces of joined B_4C after the four-point bending test. Three fracture types were observed in this study. Type I represents fractures of the B_4C base material. The fracture surface for this type was several millimeters away from the joining interlayer. Thus, joining is unrelated to fracture. Next, type Ⅱ represents fractures of the B₄C base material near the joining interlayer. The joining interlayer was not observed on fracture surfaces of this type. This makes it difficult to determine whether the joint affects the fracture. Finally, type Ⅲ represents fractures of both the B₄C base material near the joining interlayer and the joining interlayer. In this type, the joining interlayer was observed on the fracture surface, indicating that the joining interlayer affects the fracture. The relationship between joining pressure and fracture type is shown in Fig. 3. Samples prepared at 0.5 (no load) and 6 kPa tended to fracture by type Ⅲ. Samples prepared at 12.0 kPa and stabilized with a carbon jig tended to fracture by types Ⅰ or Ⅱ. Therefore, joined B₄C with a flexural strength similar to that of the B₄C base material fractured from defects in the B₄C base material. In contrast, joined B₄C with a flexural strength lower than that of the B₄C base material fractured at the joining interlayer.

Figure 4 shows SEM micrographs of the B₄C joining layer at 0.5 kPa and 12 kPa. At 12 kPa, the thickness of the joining layer was about 7 μm, which is thinner than the 10-μm thickness of the original Al sheet before joining. In addition, void-free and crack-free interfaces were observed. Boron carbide appears to be strongly bonded with Al from SEM observations. In contrast, in the absence of load (0.5 kPa), the thickness of the joining layer was about 10 μm, which is the same as that of the original Al sheet. Furthermore, some voids were observed in the joining interlayer.

Suganuma et al.[15] reported the effect of applied pressure on the strength of the joint between silicon nitride and Al. They reported that the strength of silicon nitride joined at 0 MPa was lower than that of silicon nitride joined at 0.05 MPa, and that there were many pores. In addition, the thickness of the aluminum layer was not uniform at 0 MPa, whereas it was thinner than the original Al sheet at 0.05 MPa. They concluded that microstructural differences between the samples joined at 0 MPa and 0.05 MPa influenced their strengths. Their findings are similar to the results of B₄C joining in the present study.

A schematic of the B₄C joint is illustrated in Fig. 5. Before joining, several voids apparently exist between the B₄C base material and the Al sheet because the latter is distorted. In the case of low-pressure joining (e.g., at 0.5 or 6 kPa), Al melted; however, the thickness of the joining interlayer was the same as that of the original Al sheet because the

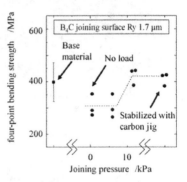

Fig. 1 Dependence of flexural strength on joining pressure

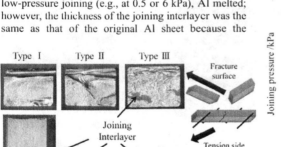

Fig. 2 Fracture surface of joined B₄C after four-point bending test

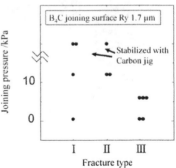

Fig.3 Relationship between joining pressure and fracture type of joined B₄C

molten Al failed to flow owing to the friction between Al and B$_4$C. Presumably, the voids in the joint that existed before joining remained after joining. Therefore, the joining area of low-pressure samples decreases owing to the voids. In turn, this presumably decreases the joining strength as well. As a result of the low joining strength, the joining interlayer was observed on the fracture surface. In contrast, in case of high-pressure joining (e.g., 12 kPa or carbon jig stabilizing), the thickness of the joining interlayer was thinner than that of the original Al sheet. This is because Al flowed owing to the outer pressure and the voids moved out of the joining interlayer according to the Al flow in vacuum. An outer pressure over 12 kPa or stabilizing with a carbon jig is important to create a dense joining interlayer and high flexural strength.

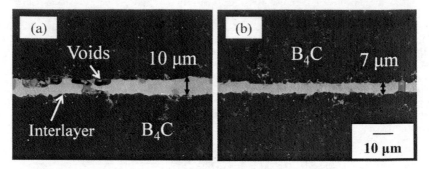

Fig. 4 SEM micrograph of B$_4$C joining interlayer under (a) 0.5 kPa (no load) (b) 12 kPa

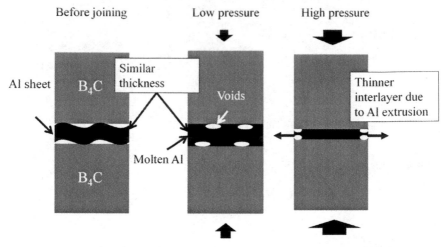

Fig. 5 Schematic of B$_4$C joining interlayer

Dependence of flexural strength on surface roughness

Next, the influence of the joining surface roughness on the strength of joined B_4C was investigated. The joining surface was polished by diamond paste or ground by a diamond wheel. The joining surface roughness Ry of the mirror-polished surface was 0.1 μm and that of the ground surface was 1.0, 1.7, or 6.1 μm. Figure 6 (a) shows the dependence of the flexural strength of joined B_4C stabilized with a carbon jig on the joining surface roughness. The four-point bending strength of joining samples with a roughness 0.1 μm (i.e., of the mirror-polished surface) was close to that of the B_4C base material. The four-point bending strength of the ground-surface joining samples with a joining surface roughness in the range of 1.0 to 1.7 μm was also similar to that of the B_4C base material. However, the flexural strength of the ground-surface joining samples with a surface roughness 6.1 μm decreased slightly.

The relationship between the joining surface roughness and the fracture type for joined B_4C stabilized with a carbon jig is shown in Fig. 6 (b). At a roughness ranging from 0.1 to 1.7 μm, samples primarily displayed type I fractures. Thus, in most samples, fractures occurred at the B_4C base material. This result indicates that the joining surface roughness in the range of 0.1 to 1.7 μm did not affect the flexural strength of joined B_4C.

SEM micrographs of the B_4C joining interlayer are illustrated in Fig. 7. Fig. 7 (a) and (b) show the joining interlayer of a sample with a mirror face of Ry 0.1 μm, whereas Fig. 7 (c) and (d) show that of a sample with a ground surface of Ry 1.7 μm. The thickness of the joining interlayer with a mirror-polished surface (Ry 0.1 μm) was about 1 μm and that with ground surface (Ry 1.7 μm) was about 5 μm. The thickness of the joining interlayer increased as the joining surface roughness increased. Moreover, little damage was observed at the interface in the mirror-surface sample, and B_4C appeared to be strongly bonded with Al. In contrast, cracks and voids were observed in the B_4C at the interface of the sample with a ground surface, although Al permeated and filled in most of these cracks and voids. In particular, Al permeated into a narrow B_4C microcrack with a width of about 20 nm and an aspect ratio of about 105, as shown in Fig. 7 (d). Al thus exhibits high permeability into B_4C cracks.

In general, the relationship between infiltration pressure (P) and contact angle (θ) is represented by the equation

$$P = 2\gamma\cos\theta/r \qquad (1)$$

where γ is the surface tension, r is the radius of the tubular defect. Based on the literature, the surface tension of Al is 914 (mN/m)[17] and the contact angle between Al and B_4C is about 20° at 1000 °C for 2 h.[18] In addition, r is about 10 nm. Therefore, P is calculated to be about 172 MPa. The molten Al presumably permeated into narrow B_4C microcracks during B_4C joining owing to this pressure. In addition, the viscosity of Al at 1000 °C is about 1 mPa.[19] This value, which is similar to that of water at room temperature, is also beneficial for the permeation of Al melt. This phenomenon of microcrack permeation by Al presumably contributes to the reduction of defects in the joining interlayer. It explains why the flexural strength of ground-surface samples with a roughness of 1.0 to 1.7 μm was similar to that of the B_4C base material and that of the mirror-surface sample.

Figure 7 (e) and (f) show an SEM micrograph of the B_4C joining interlayer of a sample with a ground surface of Ry 6.1 μm. The thickness was non-uniform and varied from 4 μm to 10 μm; in addition, some voids were observed in the joining interlayer. Al permeation and filling of B_4C microcracks at the joining interlayer can be observed as in the sample with a ground surface of Ry 1.7 μm. Presumably, voids that existed before joining could not be removed because of the roughness.

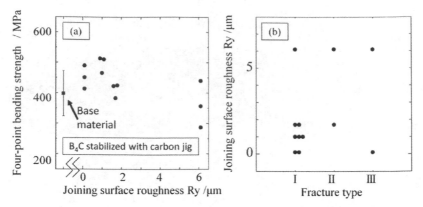

Fig. 6 (a) Dependence of flexural strength on joining surface roughness and (b) relationship between joining surface roughness and fracture type of joined B₄C

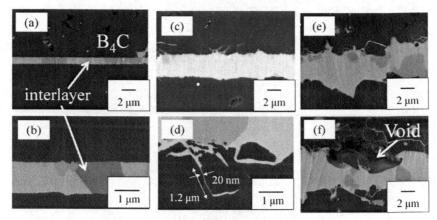

Fig. 7 SEM micrograph of B₄C joining interlayer (a) (b) mirror polished surface (Ry 0.1 μm) (c) (d) ground surface (Ry 1.7 μm) (e) (f) ground surface (Ry 6.1 μm)

CONCLUSIONS

The influence of joining pressure and surface roughness on the flexural strength of B₄C joined by Al was investigated. B₄C ceramics in which the joining surface roughness Ry varied from 0.1 to 6.1 μm were prepared. An Al sheet was sandwiched between the B₄C samples. The joining of B₄C was tested under a pressure of 0.5–12 kPa at 1000 °C for 2 h in vacuum (10^{-2}–10^{-4} Pa) or was stabilized with a carbon jig at 1000 °C for 2 h in vacuum (10^{-2}–10^{-4} Pa).

The flexural strength of B₄C joined at 12 kPa and stabilized with a carbon jig was similar to that of the B₄C base material. However, the flexural strength of B₄C joined at 0.5–6 kPa was lower than that of the B₄C base material because of the presence of voids in the interlayer. The flexural strength of joined B_4C, in which the joining surface roughness ranged from 0.1 to 1.7 μm, was similar to that of the B_4C base material. The possibility that Al permeated and filled the microcracks and voids of the B_4C at the interface is proposed.

ACKNOWLEDGEMENTS

This work was supported by the Ministry of Economy, Trade, and Industry (METI) and the New Energy and Industrial Technology Development Organization (NEDO) as part of the Innovative Development of Ceramics Production Technology for Energy Saving project.

REFERENCES

[1]F. Thevenot, Boron carbide—a comprehensive review, *J. Eur. Ceram. Soc.*, **6**, 205-225 (1990).

[2]M. A. Kuzenkova, P. S. Kislyi, B. L. Grabchuk, and N. I. Bodnaruk, Structure and properties of sintered boron-carbide, *J. Less-Common Met.*, **67**, 217-223 (1979).

[3]B. V. Cockeram, The diffusion bonding of silicon carbide and boron carbide using refractory metals, *DOE's Office of Scientific and Technical Information*, USDOE Contract No. DE-AC11-98PN38206 (1999).

[4]T. Kumazawa, T. Honda, Y. Zhou, H. Miyazaki, H. Hyuga, and Y. Yoshizawa, Pressureless sintering of boron carbide ceramics, *J. Ceram. Soc. Jpn.*, **116** [12], 1319-1321 (2008).

[5]H. Miyazaki, Y. Zhou, H. Hyuga, Y. Yoshizawa, and T. Kumazawa, Microstructure of boron carbide pressureless sintered in an Ar atmosphere containing gaseous metal species, *J. Eur. Ceram. Soc.*, **30**, 999-1005 (2010).

[6]M. Mashhadi, E. Taheri-Nassaj, V. M. Sglavo, H. Sarpoolaky, and N. Ehsani, Effect of Al addition on pressureless sintering of B_4C, *Ceram. Int.*, **35**, 831-837 (2009).

[7]J. C. Viala, J. Bouix, G. Gonzalez, and C. Esnouf, Chemical reactivity of aluminum with boron carbide, *J. Mater. Sci.*, **32**, 4559-4573 (1997).

[8]G. Arslan, F. Kara, and S. Turan, Reaction model for the boron carbide/aluminium system, *Key Eng. Mater.*, **264-268**, 1059-1062 (2004).

[9]Q. Lin, P. Shen, F. Qiu, D. Zhang, and Q. Jiang, Wetting of polycrystalline B_4C by molten Al at 1173-1473 K, *Scr. Mater.*, **60**, 960-963 (2009).

[10]A. J. Pyzik and D. R. Beaman, Al-B-C phase development and effects on mechanical properties of B_4C/Al derived composites, *J. Am. Ceram. Soc.*, **78** [2], 305-312 (1995).

[11]W. F. Du and T. Watanabe, High-toughness B_4C-AlB_{12} composites prepared by Al infiltration, *J. Eur. Ceram. Soc.*, **17**, 879-884 (1997).

[12]X. X. Xue, D. S. Wang, and Y. Zhang, Effect of Al content on properties of B_4C-AlB_{12}-Al composites, *Chin. J. Nonferrous Met.*, **19** [9], 1594-1600 (2009).

[13]K. B. Lee, H. S. Sim, S. Y. Cho, and H. Kwon, Reaction products of Al-Mg/B_4C composite fabricated by pressureless infiltration technique, *Mater. Sci. Eng., A*, **302**, 227-234 (2001).

[14]H. M. Hu, E. J. Lavernia, W. C. Harrigan, J. Kajuch, and S. R. Nutt, Microstructural investigation on B4C/Al-7093 composite, *Mater. Sci. Eng., A*, **297**, 94-104 (2001).

[15]K. Suganuma, T. Okamoto, M. Koizumi, Joining of silicon nitride to silicon nitride and to Invar alloy using an aluminium interlayer, *J. Mater. Sci.*, **22**, 1359-64 (1987).

[16]T. Nagano and F. Wakai, Relationship between surface roughness and bonding strength in superplastic diffusion bonding, *J. Ceram. Soc. Jpn.*, **102** [7], 691-4 (1994).

[17]C. J. Smithells, Metals reference book 5th ed., *Butterworths*, 1976.

[18]D. C. Halverson, A. J. Pyzik, I. A. Aksay, and W. E. Snowden, Processing of boron carbide-aluminum composites, *J. Am. Ceram. Soc.*, **72** [5], 775-80 (1989).

[19]T. Iida and R. I. L. Guthrie, The Physical Properties of Liquid Metals, *Clarendon Press*, Oxford, 1988, ISBN 0-19-856394-9.

LASER MACHINING OF MELT INFILTRATED CERAMIC MATRIX COMPOSITE

Jarmon, D. C.[1], Ojard, G.[2], and Brewer, D.[3]

[1] United Technologies Research Center, East Hartford, CT
[2] Pratt & Whitney, East Hartford, CT
[3] NASA – Langley Research Center, Hampton, VA

ABSTRACT

As interest grows in considering the use of ceramic matrix composites for critical components, the effects of different machining techniques, and the resulting machined surfaces, on strength need to be understood. This work presents the characterization of a melt infiltrated SiC/SiC composite material system machined by different methods. While a range of machining approaches were initially considered, only diamond grinding and laser machining were investigated on a series of tensile coupons. The coupons were tested for residual tensile strength, after a stressed steam exposure cycle. The data clearly differentiated the laser machined coupons as having better capability for the samples tested. These results, along with micro-structural characterization, will be presented.

INTRODUCTION

There is ever increasing interest in ceramic matrix composites (CMCs) for multiple high temperature applications [1,2]. This is shown by the interest in CMCs for gas turbine engines for turbine and combustor applications [3,4]. The reason for this interest is the high temperature capability of CMCs with increased damage tolerance present from the high strength fibers and weak fiber interface of the composite [2]. Extensive work is ongoing to characterize CMCs for various insertion opportunities. As for any material, part of the consideration for the use or insertion of the material is the machining. This is especially true for CMCs because the machining method and machining parameters can significantly affect the performance. As would be expected for a CMC material, machining would initially start with diamond grinding, since this was a standard procedure for monolithic ceramic materials. The ASTM International (ASTM) standards exist for testing of ceramic materials which include detailed machining procedures [5]. The ASTM machining procedures are used extensively for CMC materials and can be done either dry or wet (coolant) . There may be some concerns about coolant use and its effect on the weak interface.

With this background, different machining approaches were considered for the melt infiltration (MI) SiC/SiC system including: water jet, electrical discharge, and laser machining. Electrical discharge could be considered because of the presence of the conductive , metallic Si alloy phase in the MI SiC/SiC system. Laser machining was chosen due to the fact that no liquid medium was required. This eliminated any concerns about liquid effects on the weak interface. A series of samples were machined with both diamond grinding and laser machining. These samples were then tested in a steam environment under load for 100 hours. After exposure, residual properties and micro-structural characterization was performed. The results of this work will be presented.

113

PROCEDURE

Material

The material used for this study was the Melt Infiltrated SiC/SiC CMC system as noted previously. This material has been studied extensively [6,7]. The material consists of a stochiometric SiC fiber in a complex sequenced matrix. The Sylramic® fiber was fabricated by DuPont as a 10 μm diameter stochiometric SiC fiber and bundled into tows consisting of 800 fibers. The sizing applied was polyvinyl alcohol (PVA). The tow spools were then woven into a 5 harness satin (HS) balanced weave at 18 per inch (EPI). This resulted in 36% volume fraction of fibers. The fabric was then laid in graphite tooling to correspond to the final part design (flat plates for this experimental program). All the panels were manufactured from a symmetric cross ply laminate using a total of 8 plies. The graphite tooling has holes to allow the CVI deposition to occur. In the first CVI phase, the BN layer was applied to provide the weak interface (0.5 microns thick). This was followed by SiC vapor deposition around the tows. Typically, densification is done to about 30% open porosity. SiC particulates are then slurry cast into the material followed by melt infiltration of a Si alloy to arrive at a nearly full density material. The material at this time has less than 2% open porosity. Typical cross sections of this material are shown in other work by the authors [8]. The predominance of the Si phase is shown in Figure 1.

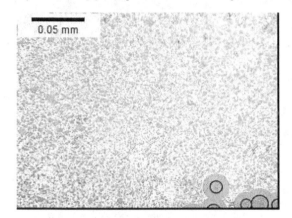

**Figure 1. Optical Image showing Silicon Matrix Phase (white)
with SiC Particulates (grey). SiC fiber coated with CVI SiC
in lower right corner**

Machining

Diamond Grinding:

The baseline for the diamond grinding data was from the existing material database as noted previously [6,7]. For this effort, there were two vendors used who nominally use the same process. The material to be machined was mounted to a fixture plate using a low temperature bonding adhesive [5]. The sample was then blanked using a slitting wheel attached to the grinding machine. Final touch up on the edges was performed with a standard surface grinder using a 320 grit resin bonded wheel. A synthetic water based coolant was used for the slitting and grinding steps [5]. After these steps the sample was available for testing and no further machining or polishing was done.

Laser Machining:
   The laser machining for this effort was done using a pulsed YAG (Yttrium, Aluminum, Garnet) laser with conditions based on experience arrived at by United Technologies Research Center (UTRC) [9]. For this effort, the linear cutting rate was set at 7.6 mm/min (0.3 in/min). It was determined that the laser can punch through the material (peck time) in 3 seconds. As the laser moved through the material, it removed 0.13 mm (0.005 in) of material, which is at least one third of the kerf for diamond slicing. A CO_2 laser was not considered for this effort. It was learned that this type of laser can heat the CMC material substantially, resulting in melting or causing reactions in ceramic materials.

Characterization/Exposure/Testing:
   After the machining processes were performed, a series of characterization efforts were undertaken. Standard micro-structural characterization was done on the machined samples: optical and scanning electron microscopy (SEM). In addition, a series of dwell-fatigue tests were performed on the samples in a 90% relative humidity environment. The dwell fatigue test was a modified creep test that was performed at an elevated temperature of 815°C (1500°F). The load was cycled on and off every two hours in order to break down any protective oxides that would form in cracks resulting from the testing [2]. For this testing, there were 4 samples from the diamond grinding (two vendors) and laser machining. (This resulted in 8 samples being diamond ground and 4 samples being laser machined. There is no effort in this paper to differentiate between the two diamond grinding vendors.) The stress level for the testing was set at 117.3 MPa (17 ksi). This level is below the offset proportional limit but above the micro-cracking stress of the material [10]. The run-out time for this series of tests was set at 100 hours. If the samples did not fail, residual tensile tests were to be done at room temperature.

RESULTS

Optical Cross Sections (Pre Test)
   Optical cross sections from the different machining approaches are shown in Figures 2 and 3. Figure 2 is representative of images from the diamond grinding effort. The images from the diamond grinding approach show cracking, possibly a result from the machining effort. The severity of the cracking, shown in Figure 2, was not seen for the full breadth of the diamond ground specimen, but minor cracking was seen in all the images taken. The optical microstructure from the laser machined specimen is shown in Figure 3. Cracking is still observed, but it does not extend as far into the material as was the case for the diamond grinding. A distinct difference can be seen in the laser machining effort as a re-cast layer formed on the surface. It is clear that this resulting layer is not uniform, but was present at all machined surfaces. The recast layer formed such that any cracks formed (as shown in Figure 3) were not surface breaking (as was seen in Figure 2 for the diamond ground samples).

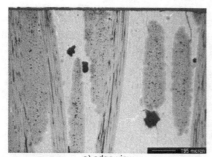
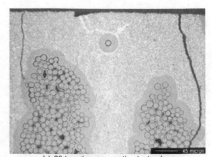

| a) edge view | b) 0° tow (cross section) at edge |

Figure 2. Optical Cross Section from Sample that was machined by Diamond Grinding

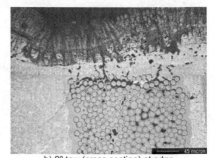

| a) edge view | b) 0° tow (cross section) at edge |

Figure 3. Optical Cross Section from Sample that was machined by Laser

Exposure (Steam)

The 12 samples machined for this effort were all tested in a 2-hour dwell fatigue test at 815°C (1500°F). This testing was not done in air, but the rig was modified to allow the chamber to hold a steam environment of 90% relative humidity. As was noted, the run-out time for this testing was set at 100 hours. All 12 samples achieved the run-out time of 100 hours. There were no failures of the samples. All samples were subsequently tensile tested at room temperature to determine their residual properties.

Tensile Testing – Residual Properties

Room temperature tensile testing was performed on the 12 samples after steam exposure. The average results of this testing are shown in Table I. Table I shows that there was no difference in the in-plane tensile modulus between the two machining methods. The laser machining shows a higher strength and strain to failure. This is also shown in the select stress-strain curves, shown in Figure 4. The average material Ultimate Tensile Strength (UTS), from diamond ground specimens without exposure, was 336 MPa, with a corresponding strain to failure of 0.0038 mm/mm [6,7]. And a corresponding modulus of 273 GPa. It is clear that the exposure (under load and steam) had reduced the material strength and strain to failure, but to differing extent depending on the machining approach.

Table I. Average Room Temperature Residual Tensile Test Results

Machining Method	Analysis	Modulus (GPa)	UTS (MPa)	Strain to Failure (mm/mm)
Diamond Grinding	Average	254.2	188.7	0.00095
	StDev	28.32	23.66	0.000143
Laser Machining	Average	258.6	270.2	0.00196
	StDev	9.07	7.46	0.000103

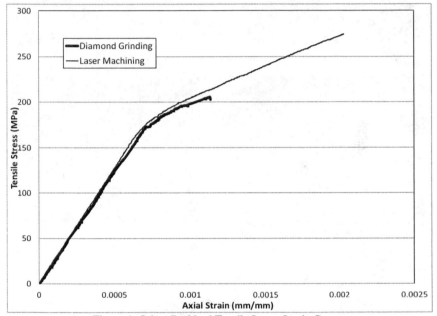

Figure 4. Select Residual Tensile Stress-Strain Curves

SEM Failure Face Analysis (Post Exposure and Residual Testing)

In order to gain additional insight into the residual stress-strain behavior, failed tensile bars were interrogated in an SEM. Figures 5 and 6 show images of samples from the different machining methods used. Figure 5 shows that there is very little fiber pullout present between the two methods and this was consistent with the low strain to failure. Higher magnification work was done (See Figure 6) and the images taken from the diamond grinding effort show that the interface has been oxidized away. The interface coating for the laser machining effort was not as affected and was consistent with the slightly higher strain to failure observed.

a) Diamond Grinding   b) Laser Machining

Figure 5. SEM images of Fracture Surface for Different Machining Methods

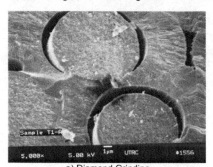
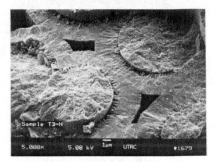

a) Diamond Grinding   b) Laser Machining

Figure 6. SEM images of Fracture Surface for Different Machining Methods Focused on Interface Region

DISCUSSION

As noted previously, the machining of CMC materials relied heavily on the machining effort of monolithic ceramics, which are known to be very flaw sensitive. Diamond grinding of CMC samples has been focused on sample mechanical performance and no known issues were raised. In addition, the early focus of CMC insertion has been in areas where machining and machining cost was not a primary focus [11].

This is the first known effort, by the authors, to look at the long term behavior of a class of CMC material in a known aggressive environment with differing machining approaches. It is clear from the work that the laser machining results in a higher strength and strain to failure (see Table I and Figure 4). This can be attributed to lower oxidation degradation as confirmed by SEM images showing that the interface was not attacked during the environmental exposure (Figure 6b).

The presence of the recast layer from laser machining (See Figure 3) appears to be the main reason for the increased residual capability. The presence of the recast layer inhibits cracks from reaching the surface. This eliminates a prime path for oxidation attack of the weak interface coating. To see if this was the case, microprobe analysis was done near the machined edge of the fully exposed tensile bars to determine if the weak interface was protected. This work is shown in Figure 7. Figure 7 clearly shows that oxygen has progressed to the weak interface, regardless of the machining approach. What is seen in Figure 7 is that the depth of penetration was greatly reduced for the laser machined sample. In addition, the recast layer appears to be an oxygen getter. This suggests that the recast layer aided in the environmental protection of the material.

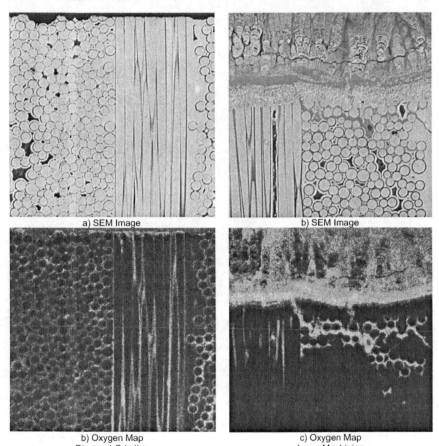

| a) SEM Image | b) SEM Image |

| b) Oxygen Map | c) Oxygen Map |
| Diamond Grinding | Laser Machining |

Figure 7. Microprobe Element Analysis for Oxygen after Exposure
(two machining cases)

CONCLUSION

The work performed has shown a clear differentiation between the two machining process, in regards to the environmental protection of the machined edges. The laser machining resulted in the formation of the recast layer which provides a barrier to oxygen ingress into the material. The clean diamond ground edge did not inhibit the ingress of oxygen. The laser machined specimens had higher residual strength, after environmental exposure, which was exhibited by a lack of attack of the fiber interface coating and oxidation of the recast layer. It was clear from these efforts that laser machining should be considered for the MI SiC/SiC material system.

FUTURE WORK

Additional work is needed to determine if faster machining rates can be used and how beveled edges could be produced. In cases where tolerances need to be kept, laser machining, followed by diamond grinding, may be an option. This would eliminate the environmental protection documented in this effort. This work also needs to be extended into other CMC systems, to see if laser machining is as effective when a metal phase is not present.

ACKNOWLEDGMENTS

Work performed under the Enabling Propulsion Materials Program, Contract NAS3-26385, Task A, David Brewer program manager.

REFERENCES

1. K.K. Chawla (1998), *Composite Materials: Science and Engineering, 2nd Ed.*, Springer, New York.
2. Brewer, D., 1999, "HSR/EPM Combustion Materials Development Program", Materials Science & Engineering, 261(1-2), pp. 284-291.
3. Brewer, D., Ojard, G. and Gibler, M., "Ceramic Matrix Composite Combustor Liner Rig Test:, ASME Turbo Expo 2000, Munich, Germany, May 8-11, 2000, ASME Paper 2000-GT-670.
4. Verrilli, M. and Ojard, G., "Evaluation of Post-Exposure Properties of SiC/SiC Combustor Liners testing in the RQL Sector Rig", Ceramic Engineering and Science Proceedings, Volume 23, Issue 3, 2002, p. 551-562.
5. ASTM C1161 "Standard Test Method for Flexural Strength of Advanced Ceramics at Ambient Temperature", American Society for Testing and Materials, West Conshohocken, PA.
6. Calomino, A., NASA-Glenn Research Center, personal communication.
7. J.A. DiCarlo, H-M. Yun, G.N. Morscher, and R.T. Bhatt, "SiC/SiC Composites for 1200oC and Above" Handbook of Ceramic Composites, Chapter 4; pp. 77-98 (Kluwer Academic; NY, NY: 2005).
8. Ojard, G., Morscher, G., Gowayed, Y., Santhosh, U., Ahmad J., Miller, R. and John, R., "Thermocouple Interactions During Testing of Melt Infiltrated Ceramic Matrix Composites", Ceramic Engineering and Science Proceedings, pp. 11-20. 2008.
9. Jarmon, D., United Technologies Research Center, unpublished research.

10. Ojard, G., Gowayed, Y., Morscher, G., Santhosh, U., Ahmad, J., Miller, R. and John, R., "Creep and Fatigue Behavior of MI SiC/SiC Composites at Temperature", Published in Ceramic Engineering and Science Proceedings, 2009.
11. Kestler, R. and Purdy, M., "SiCf/C For Aircraft Exhaust", presented at ASM International's 14th Advanced Aerospace Materials and Processes Conference, Dayton, OH, 2003.

FABRICATION OF DENDRITIC ELECTRODES FOR SOLID OXIDE FUEL CELLS BY USING MICRO STEREOLITHOGRAPHY

Naoki Komori
Graduate School of Engineering, Osaka University
Osaka, Japan

Satoko Tasaki and Soshu Kirihara
Joining and Welding Research Institute, Osaka University
Osaka, Japan

ABSTRACT

Dendritic electrodes composed of yttria stabilized zirconia (YSZ) and nickel oxide (NiO) were fabricated for solid oxide fuel cells (SOFC). These geometric structures constructed from micrometer order ceramic rods with coordination numbers of 4, 6, 8, and 12 were designed in computer graphic application. Aspect ratios of the rod lengths to diameter were valued from 0.75 to 3.00. Gaseous fluid properties and stress distributions in the dendritic electrodes were simulated and visualized by using finite element methods (FEM). The dendritic lattice with 12 coordination number was verified to exhibit the largest surface area and smooth fluid transparent characteristics. By using micro stereolithography, the optimized dendritic structure of 100 μm in lattice constant was successfully fabricated. Slurry paste of photo sensitive acrylic resin with nanometer sized YSZ and Ni particles were applied on a substrate homogeneously at 5 μm in layer thickness. Cross sectional images were exposed on the slurry surface at 2 μm in part accuracy. Through these stacking processes, lattice structures were formed. These composite precursors were dewaxed at 600 °C for 2 hs and sintered at 1400 °C for 2 hs. Microstructures were observed by using a scanning electron microscope (SEM).

INTRODUCTION

Solid oxide fuel cells (SOFCs) are promising candidate of next generation energy conversion system due to higher power density and energy efficiency. Yttria stabilized zirconia (YSZ) added nickel (Ni) has many desirable properties for SOFCs' anode, such as high electronic/ionic conductivities and chemical/mechanical stabilities at high operation temperatures. At anode site, fuel gas diffusions and electrochemical reactions on the electrode surfaces composed of YSZ/Ni/Gas triple phase boundary (TPB) proceed simultaneously. In addition, activation and diffusion overpotentials should cause to decrease SOFCs' operating voltages and energy efficiencies. Therefore, anode microstructure design is essential to achieve SOFCs' higher performances and miniaturizations. Porous YSZ-Ni anodes have been fabricated to realize large surface areas and high activation of electrode reactions[1-3]. Relationships between dispersion ratios of YSZ and Ni particles and SOFCs' output characteristics such as energy densities and overpotentials were investigated[4-6]. In addition, numerical analyses dealing with material transportations in porous structures were attempted to evaluate electrode structures[7,8]. However, it has been verified that random vacancy structures in the traditional porous anode materials obstruct fuel and produced H_2O gas diffusions and electrode reactions. In this study, SOFC anode structure was optimized through computer simulations using finite element methods (FEMs) and solid electrodes with wide surface areas and smooth fluid permeabilities were fabricated successfully by using micro stereolithography. The dendritic structures constructed from micrometer order ceramic rods propagating spatially are thought to be a desirable electrode structure, because they have large surface areas and cyclical vacancy structures. Aspect ratios of rod lengths to diameter were optimized to exhibit the maximum surface area and then streamlines, velocities and stress distributions in the dendritic structures were visualized to evaluate fluid permeabilities and mechanical strength. The dendritic structure with large surface area and smooth gaseous diffusion property are expected to activate electrode reactions and lower activation and diffusion overpotentials.

EXPERIMENTAL PROCEDURE

Firstly, electrode structure was optimized through computer simulations. The dendritic lattices of 100 μm in lattice constant constructed from micrometer order rods with coordination numbers of 4, 6, 8, and 12 were designed by using computer graphic software (Materialise Japan Co. Ltd., Japan, Majics Ver. 14). Aspect ratios of rod lengths to diameter were varied from 0.75 to 3.00 to investigate the relationships between aspect ratios and surface areas. Secondary, gaseous fluid permeabilities and stress distributions in the dendritic structures were simulated by finite volume method and FEM calculations (Cybernet Systems Co. Ltd., ANSYS Ver. 13.0). Analysis model is shown in Fig. 1. The numbers of arranged dendritic lattice unit cells were 5×5×1. Static pressure of inflow and outflow side was 1.01 and 1.00 atm, respectively. In the mechanical analysis, the bottoms of the dendritic structures were fixed. Finally, the optimized dendritic structure was formed by using micro stereolithography. A schematic illustration of the system is shown in Fig. 2. The designed model was converted into stereolithographic files of a rapid prototyping format and sliced into a series of two dimensional cross sectional image data of 5 μm in layer thickness. These image data were transferred into a micro stereolithographic equipment (D-MEC Co. Ltd., Japan, SI-C 1000). Slurry paste of photo sensitive acrylic resin with nanometer sized YSZ and Ni particles dispersions were supplied on a substrate from dispenser nozzle by air pressure and spread uniformly at 5 μm in layer thickness by a mechanical knife edge. Cross sectional images were exposed on the slurry surface at 2 μm in part accuracy. The high resolution was achieved by using digital micro-mirror device and objective lens. Through the stacking processes, three dimensional lattice structures were formed. These composite precursors were dewaxed at 600 °C for 2 hs and sintered at 1400 °C for 2 hs in air atmosphere. Microstructures of sintered sample were observed by using a scanning electron microscope (SEM).

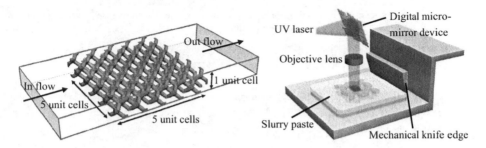

Fig. 1   A dendritic structure model for
fluid and mechanical property analysis.

Fig. 2   A schematically illustrated
micro stereolithography system.

RESULT AND DISCUSSION

  Relationships between surface areas and aspect ratios determined by rod diameters and lengths are
shown in Fig. 3. Each dendritic lattice showed the maximum surface areas when aspect ratio was 0.90,
1.17, 2.18 and 2.34, respectively. The dendritic lattice with coordination number 12 and aspect ratio
2.18 were verified to exhibit the largest surface area and the electrode texture is expected to increase
the TPB points and lower the activation potential in the electrode.

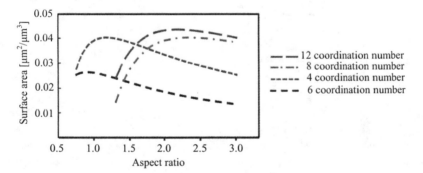

Fig. 3   Calculated variations of surface areas in dendritic structures according to aspect
ratios and coordination numbers.

  Streamlines and velocities in the dendritic structures are shown in Fig. 4. Smooth streamlines
according to cyclical vacancies were exhibited in the dendritic structure with coordination numbers of
6, 8 and 12. The prompt fuel gas flows can realize reduction of diffusion overpotentials. However,

there were rectilinear streamlines and large flow velocity differences in the dendritic structure with 6 and 8 coordination numbers. These could be causes of inhomogeneous fuel gas diffusion, forexamplebackwaters or vortexs. On the other hand, smooth and homogeneous fluid diffusion was indicated in the dendritic structure with 12 coordination number. In the SOFC anode, H_2O produced in electrode reactions must be discharged from electrode promptly to avoid lowering concentrations and partial pressures of fuel gases. A stress distribution applied by fluid flow in the dendritic structure with 12 coordination number is shown in Fig. 5. Stress concentration did not observed, and the dendritic structure can exhibit high mechanical strength by isotropic lattice structures.

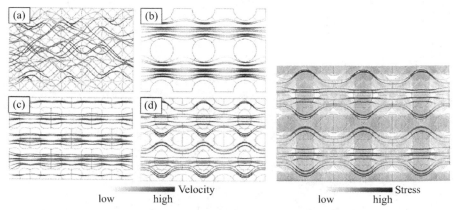

Fig. 4   Streamlines in the dendritic structure with coordination numbers of (a) 4, (b) 6, (c) 8, and (d) 12 calculated by using finite volume methods.

Fig. 5   A stress distribution in the dendritic lattices with 12 coordination number calculated by using a finite element method.

Fabricated dendritic structure composed of YSZ and nickel oxide NiO is shown in Fig. 6-(a). Micrometer order ceramic lattices with the 12 coordination number were successfully formed. Lattice constant of sintered sample was 98.5 μm. Figure 6-(b) shows the microstructure of YSZ-NiO electrode surface observed by a SEM. The YSZ and NiO particles were connected successfully and distributed homogeneously. The micro voids or pores were not observed. XRD patterns of the sintered sample are shown in Fig. 6-(c). Diffraction peaks of YSZ and NiO were shown clearly. The ideal TPB could be created in the sintered dendritic lattices.

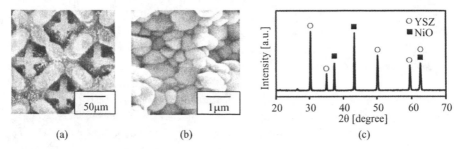

Fig. 6   A sintered dendritic structure with 12 coordination number composed of YSZ and nickel oxide (a) and A SEM image of the electrode surface (b). XRD patterns of sintered YSZ-NiO composite (c).

CONCLUSION

Dendritic electrode for solid oxide fuel cell (SOFC) was fabricated. In the optimization of anode structure, the dendritic structure with 12 coordination number showed largest surface area and smooth gaseous fluid diffusion properties. The optimized dendritic structure composed of yttria stabilized zirconia (YSZ) and nickel oxide (NiO) was successfully fabricated by using micro stereolithography. From microstructure observation by using a scanning electron microscope, YSZ and NiO particles were well connected. Fabricated dendritic electrode with large surface area and smooth fluid permeabilities is considered to activate electrode reactions effectively and lower activation and diffusion overpotentials.

REFERENCE

[1]Y. Li, Y. Xie, J. Gong, Y. Chen, Z. Zhang, Preparation of Ni/YSZ materials for SOFC anodes by buffer-solution method, Material Science and Engineering, B86 (2001), 119-122 1

[2]J.-H. Lee, H. Moon, H.-W. Lee, J. Kim, J.-D. Kim, K.-H. Yoon, Quantitative analysis of microstructure and its related electrical property of SOFC anode, Ni-YSZ cermet, Solid State Ionics, 148 (2002), 15-26

[3]K.-R. Lee, S.H. Choi, J. Kim, H.-W. Lee, J.-H. Lee, Viable image analyzing method to characterize the microstructure and the properties of the Ni/YSZ cermet anode of SOFC, Journal of Power Sources, 140 (2005), 226-234

[4]H. Koide, Y. Someya, T. Yoshida, T. Maruyama, Properties of Ni/YSZ cermet as anode for SOFC, Solid State Ionics, 132 (2000), 253-260

[5]T. Fukui, S. Ohara, M. Naito, K, Nogi, Performance and stability of SOFC anode fabricated from NiO-YSZ composite particles, Journal of Power Sources, 110 (2002), 91-95

[6]J.H. Yu, G.W. Park, S. Lee, S.K. Woo, Microstructural effects on the electrical and mechanical properties of Ni-YSZ cermet for SOFC anode, Journal of Power Sources, 163 (2007), 926-932

[7]Y. Suzue, N. Shikazono, N. Kasagi, Numercial simulation of mass transfer and electrochemical reaction for microscopic structure design of SOFC porous electrode, JSME annual meeting, 7 (2006), 181-182

[8]M. Koyama, K. Ogiya, T. Hattori, H. Fukunaga, A. Suzuki, R. Sahnoun, H. Tsuboi, N. Hatakeyama, A. Endou, H. Takaba, M. Kubo, C.A.Del Carpio, A. Miyamoto, Development of three-dimensional porous structure simulator POCO^2 for simulations of Irregular porous materials, J. Coput. Chem. Jpn., 7 (2008), 55-62

ION-EXCHANGE PROPERTIES OF NANO ZEOLITE A PREPARED BY BEAD MILLING AND POST-MILLING RECRYSTALLIZATION   METHOD

Toru Wakihara, Ryuma Ichikawa,Junichi Tatami, Katsutoshi Komeya, Takeshi Meguro
Graduate School of Environment and Information Sciences, Yokohama National University
Yokohama 240-8501, Japan   (e-mail: wakihara@ynu.ac.jp)

ABSTRACT
    A top-down approach has been applied for the preparation of nano zeolite A using a combination of bead milling and post-milling recrystallization. This gives nano zeolite A approximately 55 nm in size with high crystallinity. Zeolites with nanometer size particles have advantages over micron meter size particles for greater diffusion of ions due to their high external surface area; however, nano zeolites prepared by the above method have yet to be applied as ion exchanger.   In this study, therefore, nano zeolite A has been prepared by bead milling and subsequent recrystallization from a dilute aluminosilicate solution and effect of crystallinity and particle size on ion exchange properties such as exchange velocity and capacity were investigated.

INTRODUCTION

    Zeolites are hydrated, crystalline tectoaluminosilicate structures, formed from TO_4 tetrahedra where T indicates a tetrahedral atom, e.g. Si, Al, P, Zn, Ge, etc. Their intricate pore and channel system at sizes similar to simple molecules is the source of their immense importance in catalysis, adsorption and ion-exchange.[1] There is ongoing interest in the synthesis of nano zeolites[2,3] since zeolites with nanometer size particles allow for greater diffusion of ions and molecules into the structures and allow greater access to the internal pore sites due to their high external surface area. In general, the fabrication of nano zeolites has been achieved by a bottom-up approach, which involves controlling zeolite nucleation and crystal growth during hydrothermal synthesis. These efforts have yielded the synthesis of many types of nano zeolites.[3,4] On the other hand, we focus our attention on a top-down approach for the fabrication of nanosized zeolites. Recently, Wakihara et al.[5,6] reported on a new method for the production of nanosized zeolite powder by a top-down approach. In those papers, zeolite A and ZSM-5 (LTA and MFI type structures, respectively) were first milled to produce nanopowders by bead milling. This technique can destroy the outer portion of the zeolite framework, which lowers the micropore volume of zeolites. To remedy this, the damaged parts were recrystallized using dilute aluminosilicate or silicate solutions after bead milling. From the combined bead milling and post-milling recrystallization, nanosized zeolites with an average size less than 100 nm were obtained (schematic illustration is shown in Figure 1). Zeolites with nanosized particles should have advantags over micronsized particles for greater diffusion of ions due to their high external surface area; however, these nano zeolites have yet to be applied as ion exchanger, the target of the present study. In this study, therefore, nano zeolite A has been prepared by bead milling and subsequent recrystallization from a dilute aluminosilicate solution and effect of crystallinity and particle size on ion exchange properties such as exchange velocity and capacity were investigated.

EXPRIMENTAL PROCEDURE

Commercial zeolite A (4A, LTA type zeolite, Si/Al=1.0, Cation: Na$^+$. Tosoh Co., Japan) was used in this study. The zeolite A was milled using a bead milling apparatus (Minicer, Ashizawa Finetech Ltd., Tokyo, Japan). 60 g of zeolite A was dispersed in 350 ml of distilled water or ethanol using an ultrasonic vibrator (VCX 600, Sonic & Materials Inc., USA) and the slurry was pulverized for 10, 60 and 120 min, respectively, using zirconia beads 300 μm in diameter. An agitation speed of 3000 rpm was used to shear and exert force on

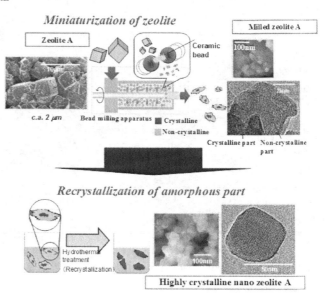

Figure 1. Schematic drawing of fabrication process of nano zeolite

the zeolite agglomerates. After milling, the slurries were dried overnight in an oven at 373 K. The recovery rate of the zeolite powder after bead milling was nearly 100%. Recrystallization of the milled zeolite A was also performed in the sample milled for 120 min using dilute aluminosilicate solution with the composition of 405 Na$_2$O : 1 Al$_2$O$_3$: 51 SiO$_2$: 29900 H$_2$O. The importance of this particular ratio is that it gives a solution nearly in equilibrium with zeolite A. This means that zeolite A is in neither macroscopic growth nor dissolution mode.[5] In these conditions poorly crystalline parts of the milled zeolite A are more easily dissolved than the more crystalline parts and tend to be recrystallized back onto the zeolite A giving a more ordered product. First, 100 ml of the aqueous solution was heated to 363 K using an oil bath. 3 g of milled zeolite was then added to the heated solution with stirring. After a period of 60 min, the slurry was centrifuged, and the supernatant liquid was decanted off. The residual solid was washed with distilled water several times. The phases present, and the morphology of the products were identified by conventional X-ray diffractometry (XRD, Multiflex, Rigaku, Tokyo, Japan), field emission scanning electron microscopy (FE-SEM, S-5200, Hitachi, Tokyo, Japan) and field emission transmission electron microscopy (FE-TEM, 2100F, JEOL Tokyo, Japan). Infra-red spectroscopy (IR) can be employed to investigate the internal structure of zeolite A. IR spectra were obtained using a IR spectrometer (JASCO Corp., Tokyo, Japan).

Zeolites obtained were in a Na form, and ion exchange property from Na^+ to Ca^{2+} was investigated by a conventional method. 1 g of the zeolite was added to 100ml of 0.035 mol/L calcium nitrate solution and stirred for prescribed period at 293 K. Then the slurry was centrifuged, and the supernatant liquid was decanted. The ion exchanged zeolites thus prepared, were filtered, washed thoroughly and dried. Then, Ca/Na ratio of the samples were measured using X-ray fluorescence spectrometry (XRF, JSX-3202, JEOL, Tokyo, Japan). Note that no changes were seen in the Si/Al ratios throughout the treatments; indicating the composition of the samples are $0.5x$Ca : $(1-x)$Na : 1.0Si : 1.0Al : 4.0O, where x=0-1.

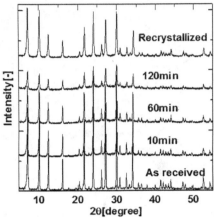

Figure 2. XRD patterns of the samples.

RESULTS AND DISCUSSION

XRD patterns of the samples are shown in Figure 2. The diffraction peaks assigned to an LTA structure in the bead milled samples showed the sample has still crystalline but the peak intensities had decreased with increasing the milling period, indicating a decrease in crystallinity by the bead milling treatment. After recrystallization, however, the crystallinity returns closely to the original level. It is worthy of note that there was almost no loss of mass during the recrystallization treatment with the

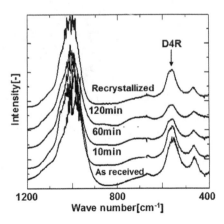

Figure 3. IR spectra of the samples.

recrystallized product having 97% of the mass of the milled starting material. It appears that selective recrystallization occurs at the poorly crystalline parts of the milled zeolite A, where damage has been caused by bead milling. The Si/Al ratio of the samples measured by XRF shows no changes, that is Si/Al=1; indicating that the material balance was maintained during the milling and recrystallization treatments. Changes in the ring connectivities were also investigated by IR as shown in Figure 3. It has been shown that bands in the 550 cm^{-1} region are sensitive to the double 4-membered rings, the framework connectivity of the ring systems present in the zeolite A. These peaks were seen in all three samples. The double 4-membered rings peaks become weakened after bead milling treatment; indicating the framework connectivity has become distorted and/or

Figure 4. SEM images of the samples.

destroyed by shearing and other forces imparted on the zeolite A. After the recrystallization, the peak becomes sharper and is similar to the as-received zeolite A. These results indicate recrystallization repairs the crystallinity of the zeolite A after the damage caused by bead milling and the framework connectivity was improved, supporting XRD results.

Typical FE-SEM images of the samples are shown in Figure 4. The as-received zeolite has smooth morphological features. After bead milling the zeolite A the morphology has changed dramatically. The as-received zeolite (average: 2 μm) became tiny particles about 50-500 nm (average 350 nm) and 30-300 nm (average 150 nm) after bead milling for 60 and 120 min, respectively. After recrystallization, the product has become sharply-defined nanoparticles approximately 30-100 nm across. The average particle size after recrystallization was estimated to be 55 nm from the FE-SEM image. TEM images before and after recrystallization are shown in Figure 5. Mixture of crystalline and non-crystalline parts was confirmed in the sample after bead milling treatment for 120 min. On the other hand, the zeolite nanoparticles were highly crystalline after post-milling recrystallization. No residual non- crystalline part or layer was present. It appears that a large number of crystallites were formed by bead milling and each of them grew uniformly during the post-milling recrystallization, resulting in nanoparticles with high crystallinity.

Figure 6 shows Ca/Na ratio of as-received, milled and recrystallized samples along with ion exchange period. It is confirmed that arrival time to the equilibrium state is dramatically shortened in the samples after bead milling treatments, indicating the zeolites with nano size allow for greater diffusion of ions and molecules and allow them easier access to the internal pore sites. Ca/Na ratio of saturated value after bead milling, however, is lower than as-received zeolite A. It is assumed that the amorphous part formed by bead milling has less or no contribution to the ion exchange. On the other hand, Ca/Na ratio of zeolite after recrystallization treatment showed the highest value as shown in Figure 7, that is, ion exchange property was improved as a result of combination of bead milling and recrystallization treatments.

CONCLUSIONS

A top-down approach has been applied for the preparation of nano zeolite A using a combination of bead milling and post-milling recrystallization. Nano zeolite A approximately 55 nm in size with high crystallinity was obtained. Effect of particle size and crystatllinity on ion exchange velocity and capacity were investigated. It is

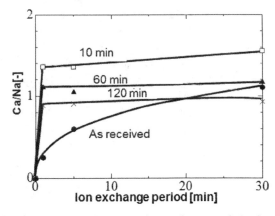

Figure 5. TEM images of the samples before and after recrystallization treatment. (a) bead milled for 120 min, (b) after recrystallization treatment

confirmed arrival time to the equilibrium state is dramatically shortened in the samples after bead milling treatments; indicating the zeolites with nano size allow for greater diffusion of ions and molecules and allow them easier access to the internal pore sites. Ca/Na ratio of saturated value after bead milling, however, is lower than as-received zeolite A. On the other hand, Ca/Na ratio of zeolite after recrystallization treatment showed the highest value, that is, ion exchange property was improved as a result of combination of bead milling and recrystallization

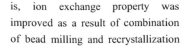

Figure 6. Ca/Na ratio versus ion exchange period of as-received and milled zeolite.

treatments.

ACKNOWLEDGEMENT

The authors would like to thank Profs. T. Tatsumi and T. Yokoi from the Tokyo Institute of Technology for FE-SEM measurements. This work has been supported partially by a "Grant for Advanced Industrial Technology Development" in 2011 from the New Energy and Industrial Technology Development Organization (NEDO) of Japan.

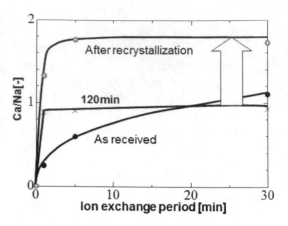

Figure 7. Ca/Na ratio versus ion exchange period of as-received, milled (120 min) and recrystallized zeolite.

REFERENCES

[1] C. S. Cundy, P. A. Cox, "The hydrothermal synthesis of zeolites: History and development from the earliest days to the present time" *Chem. Rev.*, 103, 663 (2003).

[2] L. Tosheva, V. P. Valtchev, "Nanozeolites: Synthesis, crystallization mechanism, and applications" *Chem. Mater.*, 17, 2494 (2005).

[3] W. Fan, M. A. Snyder, S. Kumar, P. S. Lee, W. C. Yoo, A. V. McCormick, R. L. Penn, A. Stein, M. Tsapatsis, "Hierarchical nanofabrication of microporous crystals with ordered mesoporosity" *Nature Mater.*, 7, 984 (2008).

[4] S. Mintova, N. H. Olson, V. Valtchev, T. Bein, "Mechanism of zeolite a nanocrystal growth from colloids at room temperature" *Science*, 283, 958 (1999).

[5] T. Wakihara, R. Ichikawa, J. Tatami, A. Endo, K. Yoshida, Y. Sasaki, K. Komeya, T. Meguro, "Bead-Milling and Postmilling Recrystallization: An Organic Template-free Methodology for the Production of Nano-zeolites" *Cryst. Growth Des.*, 11, 955 (2011).

[6] T. Wakihara, A. Ihara, S. Inagaki, J. Tatami, K. Sato, K. Komeya, T. Meguro, Y. Kubota, A. Nakahira, "Top-down tuning of nanosized ZSM-5 zeolite catalyst by bead milling and recrystallization" *Crystal Growth & Design* 11, 5153 (2011).

THE ROLE OF MILLING LIQUIDS IN PROCESSING OF METAL-CERAMIC-PRECURSOR POWDERS

Nadja Holstein[1], Katharina Wiegandt[3], Florian Holleyn[1], Jochen Kriegesmann[1], Michael R. Kunze[2], Joachim Scholz[2], Rolf Janssen[3*]

[1]Materials Science, Ceramics, University of Applied Sciences Koblenz, Höhr-Grenzhausen, Germany
[2]Institute for Integrated Sciences, University Koblenz-Landau, Koblenz, Germany
[3]Advanced Ceramics, Hamburg University of Technology, Hamburg, Germany

ABSTRACT

Reaction bonding of aluminum oxide (RBAO) and the reactive sintering of metal-alumina composites (s-MAC) are encouraging concepts for the preparation of highly dense components exhibiting specific properties. Precursor powders should be adapted for conventional powder metallurgical (PM) techniques. Thus, the focus is on the preparation of precursor powders for RBAO and s-MAC processing, exemplified by $Al-Al_2O_3$ and $Cr-Al_2O_3$ powders, respectively. For both variants, a fine grained, homogenous precursor is the essential prerequisite – particularly, if the goal for reactive synthesis routes is complete densification by pressureless sintering. The precursor powders had been wet grinded in an attrition mill. Organic milling liquids with different properties were investigated. The characterization is based on stability considerations of the milling suspensions. The comparison of the particle size distributions with the results of the sedimentation trails indicates the existence of different hard agglomerate structures build up in the milling process. According to the solvent properties the milling results can be classified not only to polar and nonpolar, but rather to polar protic, polar aprotic and nonpolar solvents. In addition, the influences of the powder characteristics on the corresponding reaction bonding or reaction sintering mechanism are examined.

INTRODUCTION

The RBAO (reaction bonding of alumina oxide) technique [1–7] and the s-MAC process (sintered metal-alumina composites) [8–13] are encouraging alternative concepts for the preparation of highly dense components exhibiting specific properties. In both routes, mixtures of metal and ceramic powders are used as starting materials (precursors). These powder mixtures are made by a wet milling process using non-aqueous milling liquids. Green bodies, prepared out of the dried and sieved powders, are thermal treated in air (RBAO) or in nitrogen (s-MAC) in order to receive alumina and metal-alumina composites, respectively. In processing RBAO, a volume expansion occurs due to the oxidation of aluminum which counteracts the sinter shrinkage. This permits the manufacture of low-to-zero shrinkage RBAO composites. Metal reinforced alumina composites (e.g. s-MAC) have many advantages compared to single-phase ceramic materials, such as high strength, high fracture toughness and good wear resistance. Furthermore, defined electrical properties [11, 14] can be adjusted by the metal content which might be an interesting tool to create structural elements with sensoric functions. The reactive sintering process is based on the redox reaction between aluminum and the oxide layer of the metal component. Only a small amount (< 1wt%) of aluminum assists sintering and enables pressureless densification.

For both variants, a fine grained, homogeneous precursor powder mixture is the essential prerequisite for the successful reaction bonding or reaction sintering process. The size of the Al particles influences the homogeneity of the microstructure and composition. Oversized Al particles are

---

* contact: janssen@tuhh.de, holstein@fh-koblenz.de

unable for a complete and consistent reaction. Further investigations of the milling process provide optimization of the reaction bonded and reaction sintered materials.

There are many studies available concerning milling of single component systems, either metal or ceramic powders. Wet milling is the widely used method of production for hard metals and oxide powders [15, 16]. In hard, brittle materials fracture occurs with minimal particle deformation and agglomeration by welding and can be described by the Griffith theory [17]. Milling studies of oxide particles usually consider variations of milling medium and milling energy [18–21]. Differences in milling efficiency are generally explained in terms of stability and viscosity [22, 23] of milling ceramic slurries. However, only the agglomeration size is influenced by the suspension stability but not the crystallite size [23, 24]. The achievable crystallite size depends on material properties and process parameters (specific energy input, size of grinding beads, etc.). For comminution of ductile materials, however, a large proportion of the energy input to the grinding mill is expended in elastic and plastic deformations so that the grinding efficiency can be considered to be 20-50% [25]. During grinding ductile metals, deformation, breaking and cold welding of powder particles are continuously repeated, leading to mechanical alloying of particles [15, 16]. These stages of different morphology are described for milling pure chromium powder [26] and aluminium powder [27]. Apart from processing of hard metals, ductile metal powders are almost exclusively prepared in a dry milling process [16, 26–28]. There are only a few studies which are concentrated on the influences of milling liquids in wet milling of pure metal powders [29]. In particular, Arias [30, 31] examined the role of milling liquids in the mechanism of comminution of ductile metals by attrition milling. The author argued that chemical reactions of the powder with the environmental solvent should play a crucial role in preventing welding and agglomeration of the particles during the milling process.

Milling ductile/brittle material combinations have received minimal attention to date, nevertheless, there are a few studies on wet milling metal ceramic powder mixtures for reaction bonding [5, 32] and reaction sintering [33, 34]. Holz et al. [5] elaborated the milling parameters (e.g. time, rotational speed and grinding beads) for $Al/ZrO_2/Al_2O_3$ powder mixtures, for milling liquids ethanol and acetone were used. Further investigations on the effects of milling liquids on the RBAO process worked up by Watson et al. [35] using ethanol, acetone and mineral spirits. It is noted that the milling liquids have a strong effect to the particle size distribution (PSD). Essl et al. [33] examined the influence of ethanol, acetone and cyclohexane for two Al/Al_2O_3 containing powder mixtures. It was observed that milling in cyclohexane resulted in the highest milling efficiency, the particle shapes were rather spherical whereas polar media (acetone and ethanol) showed a lower milling efficiency and flake-like shapes. The reason given is that in nonpolar cyclohexane the powder form agglomerates consisting of ductile Al and brittle Al_2O_3 which are assumed to enhance the milling efficiency.

Compared to the investigations performed so far, here polar protic, polar aprotic and nonpolar solvents were chosen as milling liquids as each chemical group contains different molecular structures. Using this approach it is possible to characterize both, a wide range of dielectrical constants as well as the influence of the molecular structure of the solvents. These were examined according to two different powder systems where Al is a much more ductile metal component than Cr.

EXPERIMENTAL

Powder mixtures containing 30vol% chromium, 1.6vol% aluminium and 68.4vol% aluminum oxide were used to prepare chromium-aluminum oxide composites (s-MAC). For preparing the reaction bonded aluminum oxide powder mixtures contain 45vol% aluminium and 55vol% aluminum oxide. Both powder mixtures were wet ground for 7 h in an attrition mill (PE 075 Netzsch-Feinmahltechnik, Selb, Germany) at a rotational speed of 700 rpm. Therefore, a 500 mL Al_2O_3 grinding vessel filled with 1500 g 3Y-TZP grinding beads of 3 mm diameter (TOSOH Corporation, Japan) and a 3Y-TZP agitator tool were used. The solid concentration of the milling slurries were about 8vol% with a ball to powder ratio of 15:1. During milling, the vessel was water cooled to a suspension temperature of about 18 °C. The mill slurry was sampled each hour. After milling, the powder was dried at room temperature and passed through a sieve with a mesh size of 125 µm. The particle size distribution (PSD) was controlled using light-scattering techniques (MasterSizer S, Malvern Instruments, UK). For this purpose, the dried powder samples were redispersed in ethanol followed by ultrasonic treatment. Suspension stability was characterized by determining the sedimentation behavior of the original slurry directly after milling. Sedimentation volumes were taken after 24 and 48 h of settling time in measuring cylinders. For characterization the agglomerate structures the dried powders were embedded in epoxy resin and then ground and polished with diamond paste (grain sizes: 15, 6, 3, 0.05 µm). Light microscope images from the transversal cross sections were characterized by the linear intercept method to determine the shape factor.

Table 1: Characteristics of starting powders (particle size distribution measured at TUHH)

Powder	Producer	Trade name	d_{50} [µm]
Al_2O_3	Almatis	CT 3000 SG	0.5
Al	Grimm	Al-Griess 99,7%	33.2
Cr	Alfar Aesar	99% Cr	20.6

Table 2: Characteristics of milling solvents used (manufacturer information; dielectric constant [36])

Solvent	Producer	Polarity	Chain	Boiling point [°C]	Vapor pressure [20°C/hPa]	Viscosity [mPa·s]	Dielectric constant [F/m]
n-Heptane	Carl Roth	nonpolar	linear	90-100	60-77	0.41	1.94
Cyclohexane	Riedel de Haen	nonpolar	cyclic	81	104	0.94	2.02
Isohexane	Carl Roth	nonpolar	branched	60	227	0.31	1.91
Methyl cyclo-hexane	Merck	nonpolar	cyclic, branched	101	48	0.73	2.02
Butyl acetate	Carl Roth	polar	linear	124-126	20	0.63	5.01
Benzoic acid methyl ester	Carl Roth	polar	cyclic	199	0.36	0.33	6.56
4-Methyl-2-pentanol	Sigma-Aldrich	polar	branched	132	4.9	4.1	10.90
Ethanol	Walter CMP	polar	linear	75-78	59	1.2	25.09
Acetone	Chemsolute	polar	linear	55-57	233		21.50
Hexanol	Carl Roth	polar	linear	157	1	0.59	12.50

Green bodies were prepared by uniaxial pressing 5x5x50 mm bars at 50 MPa followed by cold isostatic pressing at 160 MPa. Chromium-aluminum oxide composites (s-MAC) were sintered at 1600 °C in a graphite-heated resistance furnace in nitrogen atmosphere. The heating rate was 10 K/min and the dwell time was 1 h. For producing reaction bonded aluminum oxide (RBAO) the green bodies consisting of Al-Al$_2$O$_3$ mixtures were heat treated under oxidizing atmosphere in the temperature range of 400...800 °C and sintered at 1600 °C with a dwell time of 2 h. To determine the flexural strength in four-point bending tests the bending bars were deburred, ground and polished with diamond paste (grain sizes 15, 6 and 3 μm).

RESULTS AND DISCUSSION

Several solvents have been tested, as shown in Table 2. To describe the influence on the milling process, results considering a solvent from each chemical group are presented: ethanol (protic polar), butyl acetate (non protic polar) and heptan (nonpolar). Both the influence of the solvents on the particle size distribution (PSD) and suspension stability is exemplified using these three solvents. The effect of solvents on particle shapes is demonstrated for powder mixtures milled in ethanol and heptane. The influence of the solvents on the properties of the sintered specimens is shown for all tested milling liquids.

Looking at the particle size distributions, shown in Figure 1, milling results of both powder mixtures could be classified into three categories assigned to the solvents: polar protic (alcohols), polar aprotic (acetone/ester) and nonpolar (alkane hydrocarbons). Cr-Al$_2$O$_3$ powders milled in alcohols show a broad particle size range with a bi-modal distribution, where as in butyl acetate a narrower and almost mono-modal PSD was observed. Powder mixtures milled in nonpolar hydrocarbons show the narrowest mono-modal distributions. PSD of Al-Al$_2$O$_3$ powders show a few deviations compared to Cr containing mixtures. Al-Al$_2$O$_3$ powder milled in ethanol is narrower and almost mono modal. In the case of butyl acetate the PSD is similar to that of in ethanol milled powder mixtures. Milling powders using hydrocarbons result in a narrow mono-modal PSD. In spite of the deviations the refining results can be assigned to the chemical group affiliation of the solvents. SEM-images in Figure 2 show the selected powder mixtures milled in ethanol and heptane. All mixtures are extremely agglomerated; it can be argued that the particle size measured by light-scattering is more an agglomerate size than a grain size. Additionally, it should be noticed that the powder morphology is strongly dependent on the polarity of the milling liquid. It was observed that polar liquids cause a more flake like structure whereas nonpolar heptane affects a more spherical morphology. Especially the milled Al-Al$_2$O$_3$ powders show these variations of shape obviously due to the high ductility of the aluminum.

Images of the transverse cross-section of the milled powder particles support the statement that not only the particle sizes but also particle shapes correlate with the chemical group affiliation of the milling liquids. Selections of these images are presented in Figure 3. During the milling procedure in polar protic solvents particles tend to form flake like agglomerates as opposed to the spherical structures formed in nonpolar liquids. Powder milled in polar aprotic solvents show also flake like structures but smaller than of powders milled in polar protic solvents. For a more exact characterization of the particle shapes a shape factor has been investigated where a factor of 1 describes a spherical shape. Table 3 shows the determined shape factors for selected powder mixtures milled in heptane and ethanol. The smallest deviation from the ideal spherical structure occurred for Al-Al$_2$O$_3$ milled in heptane and the highest deviation for the same powder mixture milled in ethanol. For both mixtures the statement can be confirmed, that the particles shape depends on the polarity of the solvent, although influence of the solvents on milling Cr-Al$_2$O$_3$ is less pronounced.

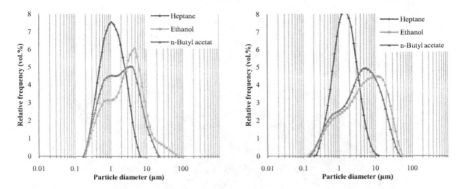

Figure 1: Particle-size distribution, left side: Cr-Al$_2$O$_3$ Powder mixtures milled for 7 h in alcohol, ketone/ester, alkane; right side: Al-Al$_2$O$_3$ powder mixtures milled for 7 h in alcohol, ketone/ester, alkane

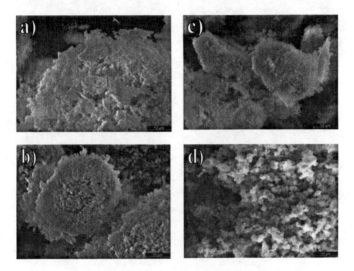

Figure 2: SEM pictures of the milled powder mixtures, left: Cr-Al$_2$O$_3$ milled in ethanol(a), and heptane(b); right: Al-Al$_2$O$_3$ milled in ethanol(c), and heptane(d)

Table 3: Shape factors for mixtures $Cr-Al_2O_3$ and $Al-Al_2O_3$ milled in ethanol and heptane

Milling liquid	$Cr-Al_2O_3$	$Al-Al_2O_3$
Heptane	2,42	1,03
Ethanol	4,76	5,24

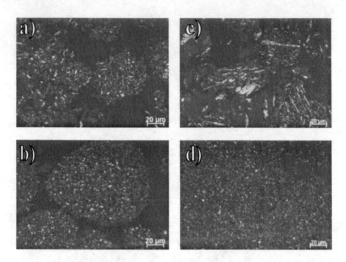

Figure 3: Light microscope images (transverse cross-section, 500x magnification) of powder particles; left: $Cr-Al_2O_3$ milled in ethanol(a) and heptane(b); right: $Al-Al_2O_3$ milled in ethanol(c) and heptane(d)

Sedimentation experiments have been conducted in order to examine the stability of the milling slurries, based on the assumption that particles in slurries with high stability achieve high packing densities which lead to small sedimentation volumes. The results are shown in Figures 4 and 5, in both powder systems ethanol shows low sedimentation volumes after 24 and 48 hours settling time, respectively. The comparison of sedimentation results indicates that ethanol slurries offer the highest stability. This is reasonable because the ability to split off agglomerates increase with higher dielectric constants [37]. Furthermore, polar liquids are able to stabilize suspensions by hydrogen bonds, which have also been suspected for $Al-Al_2O_3$ powder mixtures by Essl et al. [33]. The high sedimentation volumes of heptane slurries indicate the low stabilizing effect of nonpolar solvents. However, the influence of butyl acetate is different for Cr and Al mixtures; the $Al-Al_2O_3$ slurry seems to be more stabilized than the $Cr-Al_2O_3$ slurry. This might be due to different chemical interactions between butyl acetate and the particle surfaces of the metal components.

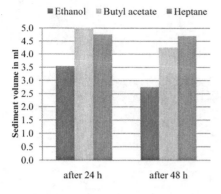

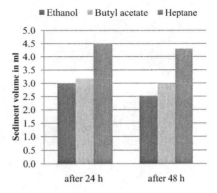

Figure 4: Sedimentation behavior of Cr-Al$_2$O$_3$ milling slurries consist of ethanol, heptanes and butylacetate after 24 and 48 hours of settling time

Figure 5: Sedimentation behavior of Al-Al$_2$O$_3$ milling slurries consist of ethanol, heptanes and butyl acetate after 24 and 48 hours of settling time

At a first glance, however, sedimentation behavior (Figures 4 and 5) seems to be contradictory to the measured particle size distributions (Figure 1), since the coarsest particles had been detected in the most stable suspensions. In order to explain this apparent contradiction the difference between the agglomerate structures existing under dynamically loaded and static systems has to be considered. Obviously, there are two types of agglomerates: the hard agglomerates which are strong enough to resist the mechanical shear stresses acting on the particles during the milling process or even ultrasonification during particle size measurement, and the soft agglomerates which are built up in static resting systems during sedimentation. This agrees with Knieke et al. [24] where a dependence of the agglomerate size on the suspension stability and on the available shear forces was observed.

In the following section the influence of the powder characteristics on the corresponding reaction bonding reaction sintering mechanisms is summarized. Figure 6 shows the values for the reactively sintered MAC specimens. The densities are in a range of 96.5 ± 0.18 %TD (for hexanol) and 97.3 ± 0.23 %TD (for cyclohexane). There is a slight tendency that higher densities resulted from powders milled in nonpolar liquids. The values for RBAO seemed to be more influenced by the milling liquids, as shown in Figure 7. Densities from about 87.3 ± 0.23 %TD (for methyl cyclohexane) up to 96.0 ± 0.15 %TD (for acetone) were determined. The extreme low densities of methyl cyclohexane specimens resulted from the fact that the precursor powder had been milled just for 2 hours; milling was canceled because of the extremely high viscosity of the milling slurry. Quite clearly, the highest densities are achieved with powders milled in ester/ketones and ethanol. However, the bending strength of acetone is about 70 MPa higher than that of ethanol specimens. Thus, it can be concluded that highest densities does not result in highest strength values. For a possible explanation further characterization of the microstructure has to be carried out.

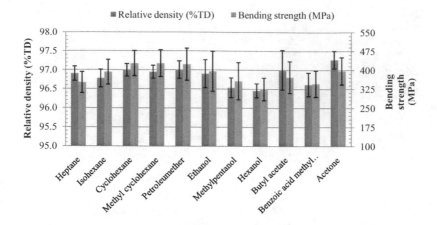

Figure 6: Sinter densities and bending strengths of the reaction sintered Cr-Al₂O₃ specimens prepared from the powders milled in different solvents

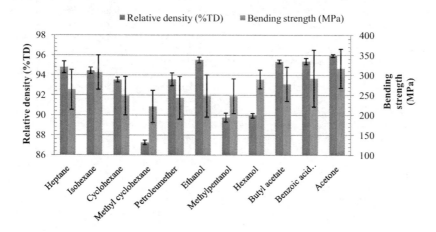

Figure 7: Sinter densities and bending strengths of the reaction bonded and sintered Al-Al₂O₃ specimens prepared from the powders milled in different solvents

CONCLUSION

For sintering process routes precursor powder properties like particle shape, reactivity and size distribution are essential. Preparation of these metal-ceramic powder mixtures implements a milling process using non-aqueous milling liquids. In addition to mechanical process parameters, chemical properties of the milling liquids are thereby crucial for the precursor quality. Previous examinations showed that the particle shapes and particle size distributions depend strongly on the polarity of the solvent used. It could be confirmed that the particle size distributions and the particle shapes depend on polarity and protic behavior, but less on the molecular structure. Cross sections of the powder particles give evidence that there are three agglomerate groups instead of two. According to the solvent properties the milling results can be classified not only to polar and nonpolar, but rather to polar protic, polar aprotic and nonpolar solvents. This is probably due to their different ability to form hydrogen bonds. Comparing the particle size distributions with the results of the sedimentation experiments indicates the existence of different hard agglomerate structures build up in the milling process. For better understanding of these formations mechanism further studies are needed.

ACKNOWLEDGEMENTS

The authors should like to thank our project partners Cerobear Bearing Technology GmbH (Herzogenrath, Germany) and INMATEC Technologies GmbH (Rheinbach, Germany) for their interest and support of our investigations. Also, the financial support grant by the German Ministry of Economy and Technology (BMWI; AiF) is gratefully acknowledged.

REFERENCES

[1] Claussen N, Le Tuyen, Wu S. Low-shrinkage Reaction-bonded Alumina. Journal of the European Ceramic Society 1989;5:29–35.

[2] Claussen N, Janssen R, Holz D. Reaction Bonding of Aluminum Oxide (RBAO) Journal of the Ceramic Society of Japan 1995;103(8):749–58.

[3] Wu S, Claussen N. Fabrication and Properties of Low-Shrinkage Reaction-Bonded Mullit. Journal of the American Ceramic Society 1991;74(10):2460–3.

[4] Wu S, Holz D, Claussen N. Mechanisms and Kinetics of Reaction-Bonded Aluminum Oxide Ceramics. Journal of the American Ceramic Society 1993;76(4):970–80.

[5] Holz D, Wu S, Scheppokat S, Claussen N. Effect of Processing Parameters on Phase and Microstructure Evolution in RBAO Ceramics. Journal of the American Ceramic Society 1994;77(10):2509–17.

[6] Suvaci E, Simkovich G, Messing GL. The Reaction-Bonded Aluminum Oxide (RBAO) Process: II, The Solid-State Oxidation of RBAO Compacts. Journal of the American Ceramic Society 2000;83(8):1845–52.

[7] Aaron JM, Chan HM, Harmer MP, Abpamano M, Caram HS. A phenomenological description of the rate of the aluminum/oxygen reaction in the reaction-bonding of alumina. Journal of the European Ceramic Society 2005;25:3413–25.

[8] Schicker S, Garcia DE, Bruhn J, Janßen R, Claussen N. Reaction Processing of Al_2O_3 Composites Containing Iron and Iron Aluminides. Journal of the American Ceramic Society 1997;80(9):2294–300.

[9] García DE, Schicker S, Janssen R, Claussen N. Nb- and Cr-Al$_2$O$_3$ Composites with Interpenetrating Networks. Journal of the European Ceramic Society 1998;18:601–5.

[10] Janssen R, Claussen N, Scheppokat S, Roeger M. Reaction Bonding and Reactive Sintering: A Way to Low Cost Manufacturing of Alumina Based Components. Materials Integration, InterMaterial 2002;15(4):75–9.

[11] Leverkoehne M, Murthy VSR, Janssen R, Claussen N. Electrical resistivity of Cr-Al$_2$O$_3$ and ZrxAly-Al$_2$O$_3$ composites with interpenetrating microstructure. Journal of the European Ceramic Society 2002;22:2149–53.

[12] Janssen R, Scheppokat S, Portue G de, Hannink R, Claussen N. Processing and Wear Behavior of Cr-Al$_2$O$_3$-ZrO$_2$ and Mo-Al$_2$O$_3$-ZrO$_2$ Composites. In: Lara-Curzio E, Readey MJ (Eds.). Ceramic Engineering and Science Proceedings: Volume 25, Issue 4. Hoboken, NJ, USA: John Wiley & Sons, Inc; 2004. pp. 203–9.

[13] Scheppokat S, Hannink R, Janssen R, Portu G de, Claussen N. Sliding wear of Cr-Al$_2$O$_3$-ZrO$_2$ and Mo-Al$_2$O$_3$-ZrO$_2$ composites. Journal of the European Ceramic Society 2005;25:837–45.

[14] Leverkoehne M, Dakskobler A, Valant M, Janssen R, Kosmac T. Cr-Al$_2$O$_3$ layered composites with a high electrical anisotropy perpared by repeated deformation processing. Journal of the European Ceramic Society 2005;25:65–72.

[15] Kuhn WE. Milling of Brittle and Ductile Materials. In: American Society for Metals (Ed.). Metals handbook. 9. ed. Metals Park, Ohio: American Soc. for Metals; 1984. pp. 56–70Powder Metallurgy; Vol. 7.

[16] Suryanarayana C. Mechanical alloying and milling. Progress in Materials Science 2001;46:1–184.

[17] Griffith AA. The Phenomena of Rupture and Flow in Solids. Philosophical Transactions of the Royal Society of London 1921;221:163–98.

[18] Kerr MC, Reed JS. Comparative Grinding Kinetics and Grinding Energy during Ball Milling and Attrition Milling. American Ceramic Society Bulletin 1992;71(12):1809–16.

[19] Padden SA, Reed JS. Grinding Kinetics and Media Wear during Attrition Milling. American Ceramic Society Bulletin 1993;72(3):101–12.

[20] Kwade A. Autogenzerkleinerung von Kalksteinen in Rührwerkmühlen [Dissertation]: Techn. Univ. Braunschweig. Shaker, 1997.

[21] Mende S. Mechanische Erzeugung von Nanopartikeln in Rührwerkskugelmühlen [Dissertation], 1. Aufl.: Techn. Univ. Braunschweig. Cuvillier, 2004.

[22] Stenger F, Peukert W. The Role of Particle Interactions on Suspension Rheology - Application on Submicron Grinding in Stirred Ball Mills. Chemical Engineering and Technology 2003;26(2):177–83.

[23] Mende S, Stenger F, Peukert W, Schwedes J. Mechanische Erzeugung und Stabilisierung von Nanopartikeln in Rührwerkskugelmühlen. Chemie Ingenieur Technik 2002;74(7):994–2000.

[24] Knieke C, Sommer M, Peukert W. Indentifying the apparent and true grinding limit. Powder Technology 2009;195:25–30.

[25] El-Schall H, Somasundaran P. Physico-Chemical Aspects of Grinding: a Review of Use of Additives. Powder Technology 1984;38:275–93.

[26] Benjamin JS, Volin TE. The Mechanism of Mechanical Alloying. Metallurgical Transactions 1974;5(8):1929–34.

[27] RODRIGUEZ JA, GALLARDO JM, HERRERA EJ. Structure and properties of attrition-milled aluminium powder. Journal of Materials Science 1997;32:3535–9.

[28] Schatt W, Wieters K, Kieback B. Pulvermetallurgie: Technologien und Werkstoffe, 2., bearb. und erw. Aufl., VDI. Springer, 2007.

[29] Rehbinder PA, Lichtman VI. Effect of surface active media on strain and rupture in solids. Electrical phenomena and solid/liquid interface 1957;3(563-582).

[30] Arias A. The Role of Chemical Reactions in the Mechanismen of Comminution of Ductile Metals into Ultrafine Powders by Grinding; NASA TN D-4862. Washington, D.C.: Lewis Research Center, National Aeronautics and Space Administration U.S.A.; 1968 [cited 2011 Jan 9].

[31] Arias A. Chemical Reactions of Metal Powders with Organic and Inorganic Liquids during Ball Milling; NASA TN D-8015. Washington, D.C.: Lewis Research Center, National Aeronautics and Space Administration U.S.A.; 1975.

[32] Suvaci E, Simkovich G, Messing GL. The Reaction-Bonded Aluminium Oxide Process: I, The Effect of Attrition Milling on the Solid-State Oxidation of Aluminium Powder. Journal of the American Ceramic Society 2000;83(2):299–305.

[33] Essl F, Bruhn J, Janssen R, Claussen N. Wet milling of Al-containing powder mixtures as precursor materials for reaction bonding of alumina (RBAO) and reaction sintering of alumina-aluminide alloys (3A). Materials Chemistry and Physics 1999;61:69–77.

[34] Schicker S, García DE, Gorlov I, Janßen R, Claussen N. Wet Milling of Fe/Al/Al$_2$O$_3$ and Fe$_2$O$_3$/Al/Al$_2$O$_3$ Powder Mixtures. Journal of the American Ceramic Society 1999;82(10):2607–12.

[35] Watson MJ, Chan HM, Harmer MP, Caram HS. Effects of Milling Liquid on the Reaction-Bonded Aluminum Oxide Process. Journal of the American Ceramic Society 1998;81(8):2053–60.

[36] Holdefer, Martin. Relative Dielektrizitätskonstante ε_r (DK-Werte) von flüssigen und festen Medien, 1999.

[37] Fowkes F. Dispersion of Ceramic Powders in Organic Media. In: Advances in Ceramics. Westerville, Ohio; 1987. pp. 411–21 .

QUANTITATIVE VALIDATION OF A MULTI-SCALE MODEL OF PYROCARBON CHEMICAL VAPOR INFILTRATION FROM PROPANE

G. L. Vignoles[1], W. Ros[1,2], G. Chollon[3], F. Langlais[3], C. Germain[2]

[1]University Bordeaux 1
LCTS – Lab. for ThermoStructural Composites
3, Allee La Boétie
F33600 PESSAC
France

[2]University Bordeaux 1
IMS – Inst. of Integration from Materials to Systems
350 Cours de la Libération
F33410 TALENCE Cedex
France

[3]CNRS
LCTS – Lab. for ThermoStructural Composites
3, Allee La Boétie
F33600 PESSAC
France

ABSTRACT

The Chemical vapor deposition (CVD) or infiltration (CVI) of pyrocarbons is a key step in the preparation of carbon-fiber reinforced carbon-matrix composites and reinforced carbon foams. Here, a simultaneous control of the deposit homogeneity in the porous medium (*i.e.* fibrous arrangement, or carbon foam) and of the pyrocarbon nanotexture is highly desirable, since it determines directly the material's thermal and mechanical properties. Pyrocarbon is usually prepared by the thermal cracking of hydrocarbons like propane or methane.

The present contribution reports an integrated multi-scale model of pyrocarbon infiltration from pure propane, featuring: (*i*) a model for gas-phase reactions during the precursor pyrolysis, (*ii*) an identification of the deposition reactions leading to specific pyrocarbon microstructures, thanks to computational fluid dynamics (CFD), (*iii*) numerical tools for multi-scale simulations of CVI and (*iv*) a finite element solver for reactor scale simulation.

The porous medium structures were acquired at various scales by X-ray CMT; previous studies on pyrocarbon CVD have allowed the construction of a consistent chemical mechanism featuring gas-phase and deposition reactions. Numerical simulations based on this data were performed and compared favorably to actual CVI experiments. Guidelines for a correct control of deposit homogeneity and microstructure are given.

INTRODUCTION

Carbon fiber reinforced carbon composites[1] are often used for high temperature applications in severe conditions. Structural parts for heat shields[2], high performance breaking systems[3] and plasma facing elements of the Tokamak[4] are made of these materials. Vibration damping[5] and shock absorption[6] properties destine carbon matrix consolidated foams to ballistic containment cases, especially at high temperature. Both these materials are commonly tailored, at industrial scale, by

Chemical vapor infiltration (CVI)[1] where a preform (a fiber arrangement or the unconsolidated foam) is to be coated by a matrix obtained through chemical cracking of a hydrocarbon precursor gas. The thermal decomposition of the precursor gas yields complex mixtures of radical and molecular species. Their eventual reaction with the perform leads to the formation of a matrix called pyrocarbon, which is a pre-graphitic form of carbon. Its nanotexture and the overall deposit homogeneity deeply reply on processing parameters such as the gas composition, pressure, temperature, the residence time or the reactor surface/volume ratio[8-14]. In their turn, final composite thermo-mechanical properties and graphitizability are directly impacted by these characteristics[15]. The costs of experimental optimization of these elaboration parameters are overruled and have triggered numerical modeling of CVI.

Previous studies in this field have given extensive knowledge of precursor gas decomposition[16-18], the evolution of geometrical characteristics[19, 20] and transport properties of a preform at various infiltration stages[21]: namely effective diffusivities in continuum or rarefied regime. Global models embodying these results have later been developed in order to optimize the final composite density as well as deposit homogeneity[22-26].

The presented numerical effort combines the prediction of both matrix nanotexture and deposition thickness. It first consists of a chemical model describing relevant homogeneous and heterogeneous reactions leading from the precursor pyrolysis to pyrocarbon formation. A multi-scale infiltration model[27, 28] is present for computation of transport and geometrical properties of the preform at different description levels. Their data is integrated to a reactor scale model for the simulation of the precursor decomposition, species transport, preform morphology evolution and matrix nanotexture.

Many CVI models are based on ideal descriptions of the fiber arrangement. However, modern imaging techniques such as X-ray computed micro-tomography (CMT) are capable of acquiring accurate 3D images of the material structure. Our infiltration model makes use of high and low-resolution images to compute effective properties at different levels of preform description. This tool works on 3D representations as can be provided by X-ray CMT but it can also handle computer generated images of idealized materials.

This paper first introduces the experimental results obtained on foam and carbon fiber arrangements. Our numerical strategy for the validation of the numerical tool is based on these datasets and introduced. A step by step presentation of this methodology, applied to the preparation of carbon reinforced composites and foams, is then detailed. Finally computational infiltrations are compared to the experimental observations and guidelines for a correct control of deposit homogeneity and pyrocarbon microstructure are given.

EXPERIMENTAL

The studied preforms were vitreous carbon foams with 60ppi pore size and carbon fiber arrangements featuring a "2.5D" reinforcement architecture (stacking of woven plies stitched together). All samples were cylindrical with 10mm in diameter and height. Their porosity, determined by the Archimedes method, was equal to 98% for the foams and 82% for the fiber arrangements. The experimental setup consists in a low-pressure hot wall tubular CVD reactor, previously described by Langlais et al..[29]. The tube diameter was 34mm and the hot zone size (defined with a 10K precision) is 30mm. The precursor gas was propane with residence times varying from 0.065 to 3s. The applied pressure ranged from 0.5 to 5 kPa whilst the temperature reached 1323K in the case of foams (F) and 1223K for the fiber arrangements (FA).

Post infiltration results and processing conditions are compiled in Table 1. The deposit thickness distribution was measured by optical microscopy. The identification of the nanotexture of each infiltrated sample was accomplished by polarized light Raman microscopy[30]. Deposition of three distinct laminar pyrocarbon nanotextures was performed. Their formation is linked to the process temperature and the residence time. When increased, one successively obtains rough laminar (RL), smooth laminar (SL) and regenerative laminar (ReL) carbon matrix.

The infiltration model makes use of 3D images of the preform at microscopic and macroscopic description levels. This is unnecessary in the case of foams as they are very porous materials where scale effects are inexistant. However, previous studies have shown that the studied fiber arrangements exhibit a bi-modal pore structure: one at the fiber scale and a second at the tow scale[31]. Both are accessible to X-ray computerized tomography but not with the same emission source. Detailed scans featuring distinct fiber bundles are only possible by using a synchrotron facility. Therefore the microscopic scale images of the fibrous preform were acquired at ESRF ID19 beamline. Effective pixel size is 0.7 μm and reconstruction of the radiographs was performed by filtered back–projection[32]. Macroscopic scale acquisitions of a foam sample and the fiber arrangement have been monitored by X-ray computerized tomography (Phoenix X-ray Nanotome). Pixel size after reconstruction was equal to 5μm in the former and equal to 25μm in the latter.

The numerical tool used for fiber scale computations requires binary representations of the image. A specific segmentation treatment developed by Vignoles[33] was used to separate the fibers for the porosity and apply a grey scale threshold.

Table 1. Experimental results: deposit thickness and matrix nanotexture.

Expt #	Reaction rates ($\mu g.cm^{-2}.min^{-1}$)	Pressure (kPa)	Residence time (s)	Duration	Thickness (μm)		Nanotexture	
					Surface	Heart	Surface	Heart
FA5-1	2.1	5	0.1	28h30min	12	7	SL	RL
FA5-2	1.7	5	0.35	18h	6	3	SL	SL
FA5-3	1.4	5	3	15h	13	3	ReL	SL
FA2-3	3.8	2	3	14h30min	4	1	ReL	SL
FA0.5-1	0.9	0.5	0.065	22h	4	1	RL	RL
F1	4.9	5	0.2	19h	25		SL	
F2	22.3	5	3	11h55min	37		ReL	

MODELING

The modeling strategy is decomposed into four steps as shown in Figure 1. Construction of the final reactor scale tool requires the outputs produced by the first three numerical models. Firstly, it must feature a semi-detailed mechanism for propane pyrolysis[11]. There are 41 species (from C1: CH_4, CH_3, etc... to C10: $C_{10}H_8$) and 133 reactions in this mechanism. In order to perform subsequent CVI computations, the reaction scheme is simplified, with a few lumped species and apparent reactions[34]. This scheme is listed in Table 2.

Using the CANTERA software[35], the applied pressure and temperature are imposed to the gas phase and species concentrations are computed. The gas-phase maturation phenomenon is evidenced: the initial decomposition of propane gives first-generation species, most in C1 (methyl radical) and C2 (ethane, ethene); then, later acetylene and benzene appear; finally, higher molecular-weight species such as naphthalene occur. Homogeneous reaction rates are identified by the combination of this data and the simplified scheme expressions.

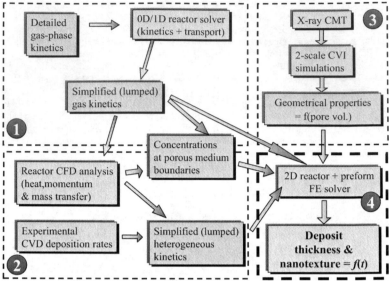

Figure 1. Flowchart of the multi-scale resolution strategy.

The incomplete kinetic model is then introduced into the FLUENT computational fluid dynamics (CFD) solver and a mass, heat and momentum analysis in a 2D axisymmetrical reactor is performed. The heterogeneous reaction rates are fitted by comparing numerically obtained deposition rates, for a given operating condition, with results from CVD experiments. When doing so favorably, concentration profiles of each species at the boundaries of the porous medium are collected.

Table 2. Simplified kinetic scheme

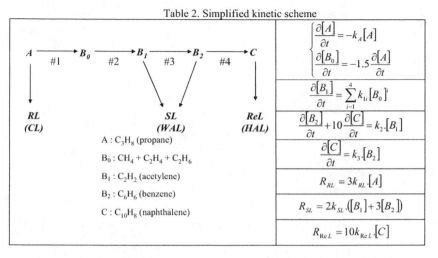

$A \xrightarrow{\#1} B_0 \xrightarrow{\#2} B_1 \xrightarrow{\#3} B_2 \xrightarrow{\#4} C$	$\begin{cases} \dfrac{\partial[A]}{\partial t} = -k_A[A] \\ \dfrac{\partial[B_0]}{\partial t} = -1.5\dfrac{\partial[A]}{\partial t} \end{cases}$

The production rate and kinetic equations are:

$$\frac{\partial[B_1]}{\partial t} = \sum_{i=1}^{4} k_{1i}[B_0]^i$$

$$\frac{\partial[B_2]}{\partial t} + 10\frac{\partial[C]}{\partial t} = k_2.[B_1]$$

$$\frac{\partial[C]}{\partial t} = k_3.[B_2]$$

$$R_{RL} = 3k_{RL}.[A]$$

$$R_{SL} = 2k_{SL}.([B_1] + 3[B_2])$$

$$R_{ReL} = 10k_{ReL}.[C]$$

RL (CL)   SL (WAL)   ReL (HAL)

A : C_3H_8 (propane)

B_0 : $CH_4 + C_2H_4 + C_2H_6$

B_1 : C_2H_2 (acetylene)

B_2 : C_6H_6 (benzene)

C : $C_{10}H_8$ (naphthalene)

The complete kinetic scheme may then be introduced in the final reactor scale 2D axisymmetrical mass balance solver for preform consolidation. The gas transport may rely on viscous flow, ordinary (binary) diffusion, and rarefied gas flow (Knudsen diffusion)[36]. In the present case, since the gases may freely flow around the porous sample, no appreciable pressure buildup appears: consequently, viscous flow is rather negligible with respect to the other transport modalities, as usual in isothermal CVI. The equations are, noting R_i the production rate of species i:

$$\frac{\partial[i]}{\partial t} + \nabla.\left(-D_{i\,por}\nabla.[i]\right) = R_i \tag{1}$$

Prior to simulation, the effective diffusivities D_{ipor} and the internal surface area σ_v, given as functions of the pore volume fraction ε, have to be determined. In this goal, the 3D CMT images and the infiltration model have been used. Deposit growth has been simulated in the high resolution scans by using a Monte Carlo random walk algorithm dedicated to fiber scale computations[27]. With this same tool, blocks representing different infiltration stages have been processed for the determination of the internal surface area and of the effective binary and Knudsen diffusivities[20, 31, 37]. The determined laws were then inserted into the composite scale algorithm[28] for similar calculations. The resulting macroscopic scale laws are then inserted into the final reactor scale solver.

The calculations in this final step of our strategy are performed using the Flex PDE commercial finite element code[38]. The temperature is assumed constant throughout the preform, since it lies in the reactor hot zone and the gas flow rate is moderate. The concentrations are given as a function of the height in the hot zone and are not prescribed directly as input values at the preform sample boundaries. Instead, a Fourier-like boundary condition is given:

$$\nabla.\left(-D_{i\,por}\nabla.[i]\right) = \frac{D_i^g}{\delta}.\left([i] - [i_0]\right) \tag{2}$$

This is related to the fact that the preform is so reactive that it is able to deplete the reac-tant concentration around it on some boundary layer thickness δ. In our case δ has been fixed to the difference between the reactor radius and the preform radius. The effective diffusion coefficient in the porous medium is given by:

$$D_i^{por} = \frac{\varepsilon}{\eta(P)}\left(\frac{1}{1+Kn}\right)D_i^{free}$$ (3)

, where the tortuosity η has been directly obtained as a function of pressure from the composite-scale random walk simulator. Then, resolution of the gas mass balances yields the gas partial pressures and deposition rate fields in the sample, as well as the effective gas fluxes at the preform boundaries. The time evolution of the sample density is given by solving the solid mass balance equations, using the values of the deposition rates computed before.

RESULTS AND DISCUSSION

Numerical simulations following the described methodology were performed for each preform and for the same experimental parameters. Table 3 proves that, whatever the physic-chemical conditions, agreement between experimental and numerical results is obtained.

The predominant nanotexture in each area is defined by studying the ratio of the deposition rates. In the case of sample FA5-3, figure 2 confirms that ReL pyrocarbon is present at the surface of the composites whereas the proportion of SL pyrocarbon is higher at the center of the material.

Table 3. Experimental and numerical results: deposit thickness and matrix nanotexture.

Expt #	Pressure (kPa)	Residence time (s)	Duration	Thickness (μm)		Nanotexture	
				Exp	Num	Exp	Num
FA5-1	5	0.1	28h30min	12-7	12-5.5	SL-RL	SL-RL
FA5-2	5	0.35	18h	6-3	6-3.5	SL	SL
FA5-3	5	3	15h	13-3	13-4.5	ReL-SL	ReL-SL
FA2-3	2	3	14h30min	4-1	4-2.5	ReL-SL	ReL-SL
FA0.5-1	0.5	0.065	22h	4-1	4.5-1.5	RL	RL
F1	5	0.2	19h	25	26	SL	SL
F2	5	3	11h55min	37	37	ReL	ReL

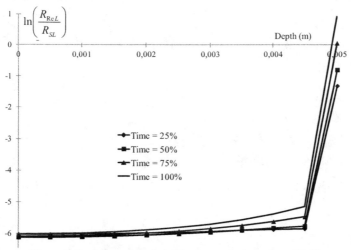

Figure 2. Evolution of the deposition rate ratio for preform FA5-3

Further analysis of these results can identify the densification scenario and measure the influence of the processing parameters. Figure 3 shows the temporal evolution of the deposition gradient in the fibrous preforms and figure 4 displays the evolution of the Thiele modulus ϕ. This indicator reflects the competition between gas diffusion and chemical reaction[39]. Important values of this dimensionless number highlight unfavorable infiltration conditions. It is defined by the internal surface area, the heterogeneous chemical reaction rate k_{het}, the characteristic length L and the effective diffusion coefficient:

$$\phi = L.\sqrt{\frac{\sigma_v.k_{het}}{D_i^{por}}} \tag{4}$$

Samples FA5-1, FA5-2 and F5-3 were processed under the same pressure but different operating residence times. The increase of this quantity favors deposition at the surface of the preform and enhances obstruction of its center to gas access. This is confirmed by the ascending rate of the outer Thiele modulus and its relatively stable inner value.

Samples FA5-1 and FA0.5-1 were prepared for similar values of the residence time. The lower pressure applied to the latter induces a weak deposition rates but delays the porous medium obstruction and prolongs ideal infiltration conditions.

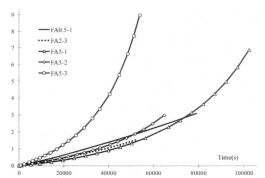

Figure 3. Evolution of infiltration gradients within the fibrous arrangements

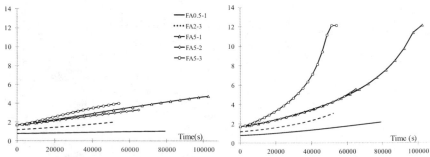

Figure 4. Evolution of the Thiele modulus. Left: Heart. Right: Surface

At a given temperature, matrix properties are a result of a compromise between the pressure and the residence time. The nanotexture can be controlled through the latter: weak and important values favor formation of highly anisotropic pyrocarbons yielding better mechanical properties. However this same quantity accelerates deposition at the surface of the preform, quickly obstructing its center to the gas phase. This phenomenon can be counterbalanced by lowering the pressure, therefore reducing the deposition rates and extending the infiltration duration.

CONCLUSION AND OUTLOOK
This work focused on the modeling of the chemical vapor infiltration of pyrocarbon from propane. The presented tool first embodies a simplified kinetic mechanism leading from the precursor decomposition to the heterogeneous reactions. It also comprises an infiltration model for multi-scale computation of geometrical and transport properties. These particular computations are based on 3D CMT scans of the preforms for accurate representation of the porous structure. The resulting data from these models is incorporated into a reactor scale finite element solver and anticipates, for a specific set of operating conditions, matrix thickness and nanotexture.

The numerical infiltrations produced on carbon fiber arrangements and foams matched with experimental observations. Studies of the Thiele modulus and the densification gradient singled out the influence of parameters and identified the morphological evolution of the porous medium. Guidelines for a correct control of matrix homogeneity and pyrocarbon nanotexture have also been produced.

Ongoing efforts consist in incorporating chemical mechanisms of other precursors: methane for example. Another objective is to adapt the presented numerical strategy to other matrix systems, such as Si-B-C.

ACKNOWLEDGEMENTS

This work has been funded by the French Ministry of higher Education and Safran – Snecma Propulsion Solide through a Ph. D. grant to W.R. The authors also wish to thank Dr C. Descamps (Safran – Snecma Propulsion Solide) for fruitful discussion.

REFERENCES

[1]G. Savage, *Carbon/Carbon composites*. Chapman & Hall. London; 1993.

[2]R. Kochendorfer, *Ceramic Matrix Composites - From Space to Earth: The Move from Prototype to Serial Production in Ceramic matrix composites*. Ceram Eng Sci Proc 2001;22:11-22.

[3]E. Fitzer and L.M. Manocha, *Carbon reinforcements and carbon/carbon composites*. Berlin: Springer; 1998.

[4]H.C. Mantz, D.A. Bowers, F.R. Williams and M.A. Witten, *A carbon-carbon panel design concept for the inboard limiter of the Compact Ignition Tokamak (CIT)*. Procs of the IEEE 13th Symposium on Fusion Engineering 1990;2:947-950.

[5]E. Bruneton, C. Tallaron, N. Gras-Naulin and A. Coscullela, *Evolution of the structure and mechanical behavior of a carbon foam at very high temperatures*. Carbon **40**, 1919 (2002).

[6]D.T. Queheillalt, Y. Katsumura and H.N.G Wadley, *Synthesis of stochastic open cell Ni-based foams*. Scr. Mater **50**,313 (2004).

[7]R. Naslain and F. Langlais, *Fundamental and practical aspects of the chemical vapor infiltration of porous substrates*. High Temp. Sci. **27**, 221 (1990).

[8]R.J. Diefendorf, *Reactivity of solids*. New York: Wiley & Sons; 1969.

[9]M.L. Lieberman and H.O. Pierson, *The chemical vapor deposition of carbon on carbon fibers*. Carbon **12**, 233 (1974).

[10]B. Reznik and K.J. Hüttinger, *On the terminology of pyrolytic carbon*. Carbon **40**, 617 (2002).

[11]C. Descamps, G.L. Vignoles, O. Féron, F. Langlais and J. Lavenac, *Correlation between homogeneous propane pyrolysis and pyrocarbon deposition*. J. Electrochem. Soc. **148**, C695 (2001).

[12]N. Reuge, G.L. Vignoles, H. LePoche and F. Langlais, *Modelling of pyrocarbon chemical vapor infiltration*. Adv. Sci. Technol. **36**, 259 (2002).

[13]G.L. Vignoles, F. Langlais, C. Descamps, A. Mouchon, H. LePoche, N. Bertrand and N. Reuge, *CVD and CVI of pyrocarbon from various precursors*. Surf. Coat. Technol. **188**, 241 (2004).

[14]I. Golecki, *Rapid vapor-phase densification of refractory composites*. Mater. Sci. Eng. **R20**, 37 (1997).

[15]A. Oberlin, *Carbonization and graphitization*. Carbon **22**, 521 (1984).

[16]C.Y. Tsai, S.B. Desu and C.C. Chiu, *Kinetic Study of Silicon Carbide. Deposited from Methyltrichlorosilane Precursor*. J. Mater. Res. **9**, 104 (1994).

[17]G.D. Papasouliotis and S.V. Sotirchos, *Heterogeneous kinetics of the chemical vapor deposition of silicon carbide from methyltrichlorosilane*. J. Electrochem. Soc. **141**, 1599 (1994).

[18]A. Li and O. Deutschmann, *Transient modeling of chemical vapor infiltration of methane using multi-step reaction and deposition models*. Chem. Eng. Sci. **62**, 4976 (2007).

[19]M.M. Tomadakis and S.V. Sotirchos, *Effect of fiber orientation and overlapping on Knudsen, transition and ordinary diffusion regime diffusion in fibrous structure.* Mater. Res. Soc. Symp. Proc. **250**, 221 (1992).

[20]G.L. Vignoles, O. Coindreau, A. Ahmadi and D. Bernard, *Assessment of geometrical and transport properties of a fibrous C/C composite preform as digitized by X-ray computerized microtomography: Part II. Heat and gas transport properties.* J. Mater. Res. **22**, 1537 (2007).

[21]J.Y. Ofori and S.V. Sotirchos, *Structural model effects on the prediction of CVI models.* J. Electrochem. Soc. **143**, 1962 (1996).

[22]T.L. Starr and A.W. Smith, *Advances in modeling of the forced chemical vapor infiltration process.* Mat. Res. Soc. Symp. Proc. **250**, 207 (1992).

[23]P. McAllister and E.E. Wolf, *Simulation of a multiple substrates reactor for chemical vapor infiltration of pyrolytic carbon with carbon-carbon composites.* AIChE J. **39**, 1196 (1993).

[24]G.L. Vignoles, C. Descamps and N. Reuge, *Interaction between a reactive preform and the surrounding gas-phase during CVI.* J. Phys. IV **10**, Pr2 (2000).

[25]N. Reuge and G.L. Vignoles, *Modelling of isobaric-isothermal chemical vapor infiltration: effects of reactor control parameters on a densification.* J. Mater. Proc. Technol. **166**, 15 (2005).

[26]J. Ibrahim and S. Paolucci, *Transient solution of chemical vapor infiltration/deposition in a reactor.* Carbon **49**, 915 (2010).

[27]G.L. Vignoles, W. Ros, C. Mulat , O. Coindreau and C. Germain, *Pearson random walk algorithms for fiber-scale modeling of chemical vapor infiltration.* Comp. Mater. Sci. **50**,1157 (2011).

[28]G.L. Vignoles, W. Ros, I. Szelengowicz and C. Germain, *A brownian motion algorithm for the tow scale simulation of chemical vapor infiltration.* Comp. Mater. Sci. **50**, 1871 (2011).

[29]F. Langlais, H. Le Poche, J. Lavenac and O. Féron, *Multiple experimental investigation for understanding CVD mechanism : example of laminar pyrocarbon deposition.* Electrochem. Soc. Proc. **2005**, 73 (2005).

[30]J.-M. Vallerot, X. Bourrat, A. Mouchon and G. Chollon, *Quantitative structural and textural assessment of laminar pyrocarbons through Raman spectroscopy, electron diffraction and few other techniques.* Carbon **44**, 1833 (2006).

[31]O. Coindreau and G.L. Vignoles, *Assessment of geometrical and transport properties of a fibrous C/C composite preform using X-ray computerized micro-tomography: Part I. Image acquisition and geometrical properties.* J. Mater. Res. **20**, 2328 (2005).

[32]A.C Kak and M. Slaney, *Principles of computerized tomographic imaging.* Classics in Applied Mathematics **33**, SIAM, Philadelphia, 2001.

[33]G.L. Vignoles, *Image segmentation for phase-contrast hard X-ray CMT of C/C composites.* Carbon **39**, 167 (2001).

[34]G.L. Vignoles, C. Gaborieau, S. Delettrez, G. Chollon and F. Langlais, *Reinforced carbon foams prepared by chemical vapor infiltration: A process modeling approach.* Surf. Coat. Technol. **203**, 510 (2008).

[35]D.G. Goodwin, *An open-source, extensible software suite for CVD process simulation*, ECS Proc, 155 (2003).

[36]J.Y Ofori and S.V. Sotirchos, *Multicomponent mass transport in CVI.* Ind. Eng. Chem. Res. **35**, 1275 (1996).

[37]G.L. Vignoles, *Modelling binary, Knudsen and transition regime diffusion inside complex porous media.* J. de Phys. IV **C5**, 159 (1995).

[38]FlexPDE software, www.pdesolutions.com.

[39]E.W. Thiele, *Relation between catalytic activity and size of particle.* Ind. Eng. Chem. **31**, 916 (1939).

Table 4. Meaning and units of the symbols used in the model.

Symbol	Meaning	Unit
D_i^{por}	Effective diffusion coefficient	$m^2.s^{-1}$
D_i^{free}	Diffusion coefficient in the free media	$m^2.s^{-1}$
k_{het}	Heterogeneous reaction rate	$m.s^{-1}$
Kn	Knudsen number	-
L	Characteristic length	m
R	Production rate of species	$mol.m^{-3}.s^{-1}$
δ	Boundary layer thickness	m
η	Tortuosity	-
ϕ	Thiele modulus	-
σ_v	Internal surface area	m^{-1}

NUMERICAL ANALYSIS OF FRACTURE BEHAVIOR IN ANISOTROPIC MICROSTRUCTURES

Hisashi Serizawa
Joining and Welding Research Institute, Osaka University
11-1 Mihogaoka, Ibaraki, Osaka 567-0047, Japan

Seigo Tomiyama and Tsuyoshi Hajima
Graduate School of Engineering, Osaka University
2-1 Yamadaoka, Suita, Osaka 565-0871, Japan

Hidekazu Murakawa
Joining and Welding Research Institute, Osaka University
11-1 Mihogaoka, Ibaraki, Osaka 567-0047, Japan

ABSTRACT
          In order to examine the microstructural fracture behavior in the advanced multifunctional materials, a finite element method with the interface element was developed, where the anisotropy of grain was modeled by the ordinary finite element while both the opening and shear deformations at grain boundary were demonstrated by the interface element.   By using virtual polycrystalline models obtained through Voronoi tessellations, the applicability of this method was studied.   As the results assuming the influence of grain orientation on the grain boundary, it was found that the anisotropic mechanical property of grain boundary (interaction between opening and slipping deformations) would be a dominant factor of the fracture process.   Also, by employing the theory of crystal plasticity for the mechanical property of grain, it was revealed that the stress concentrations caused by both the mismatch between neighbor grains and the slipping at grain boundary could be demonstrated by using this method.   Finally, it can be concluded that this method would be a useful tool for examining microstructural fracture behavior.

INTRODUCTION
          Various types of advanced multifunctional materials have been developed by controlling their microstructure precisely where these microstructures show anisotropy.   As for the practical use of these materials, it is necessary to predict not only their microstructural deformation but also fracture behavior since their mechanical properties should be estimated theoretically from the view point of structural design using these advanced materials.   Although the finite element method (FEM) has been generally used for analyzing the fracture behavior of anisotropic materials, only elastic-plastic deformations were mainly discussed in the most of previous studies because the interfacial behavior such as debonding & slipping at grain boundary cannot be demonstrated by the ordinary finite element.
          The authors have been developed the interface element as one of the methods to model the interfacial behavior directly and the various fracture behavior, which are for examples the crack propagation behavior in a plate with a center crack, the peeling process of resin coated on the steel plate and the fracture behavior of ceramic composite joints, could be demonstrated by using FEM with the interface element[1-4].   Also, the applicability of interface element for the microstructural elastic-plastic fracture behavior was revealed by using a two-dimensional virtual polycrystalline model obtained through Voronoi tessellations, where the mechanical properties of grain was assumed to be isotropic[5].   However, the mechanical properties of grain are generally anisotropic due to the grain orientation and this grain orientation also would affect the debonding & slipping at grain boundary.
          So, in this research, the anisotropic microstructural fracture behavior was examined through the finite element analyses with interface element by using the virtual polycrystalline models.   Where the anisotropic deformation of grain due to grain orientation was modeled by the ordinary finite

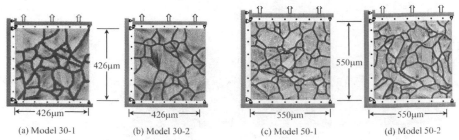

(a) Model 30-1   (b) Model 30-2   (c) Model 50-1   (d) Model 50-2

Fig. 1   Virtual polycrystalline models obtained through Voronoi tessellations.

element, while the mechanical properties at grain boundary was assumed to be affected by misorientation between neighbor grains.   Also, by employing the theory of crystal plasticity for the mechanical property of grain, the anisotropic microstructural deformations were studied by using FEM with the interface element.

ANALYSIS METHOD

Models for Analysis
        Since there have been many experimental studies about the microstructural fracture behavior in steels, the materials studied in this research was set to α-Fe and the virtual polycrystalline models of steel were created by using Voronoi tessellations where the grain size, aspect ratio of grain and grain orientations were assumed to follow the lognormal, standard and random distributions, respectively according to the microscopic observations[6].   Figure 1 shows four kinds of finite element models based on the virtual microstructures, where Model 30 and 50 consist of 30 and 50 grains, respectively. The size of Model 30 and 50 was set to 426 x 426 and 550 x 550 μm for assuming the average diameter of grain as 90 μm. The interface elements were arranged along all grain boundaries.   The mechanical boundary conditions were also shown in Fig. 1.

Interface Element
        Essentially, the interface element is the distributed nonlinear spring existing between surfaces forming the interface or the potential crack surfaces as shown by Fig. 2.   The relation between the opening of the interface δ and the bonding stress σ is shown in Fig. 3.   When the opening δ is small, the bonding between two surfaces is maintained.   As the opening δ increases, the bonding stress σ increases till it becomes the maximum value σ_{cr}.   With further increase of δ, the bonding strength is rapidly lost and the surfaces are considered to be separated completely.   Such interaction between the surfaces can be described by the interface potential.   There are rather wide choices for such potential. The authors employed the Lennard-Jones type potential because it explicitly involves the surface energy γ which is necessary to form new surfaces.   Thus, the surface potential per unit surface area ϕ can be defined by the following equation.

$$\phi(\delta_n, \delta_t) \equiv \phi_a(\delta_n, \delta_t) + \phi_b(\delta_n) \tag{1}$$

$$\phi_a(\delta_n, \delta_t) = 2\gamma \cdot \left\{ \left(\frac{r_0}{r_0 + \delta} \right)^{2N} - 2 \cdot \left(\frac{r_0}{r_0 + \delta} \right)^{N} \right\}, \quad \delta = \sqrt{\delta_n^2 + A \cdot \delta_t^2} \tag{2}$$

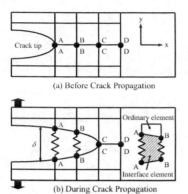

(a) Before Crack Propagation

(b) During Crack Propagation

Fig. 2　Representation of crack growth using interface element.

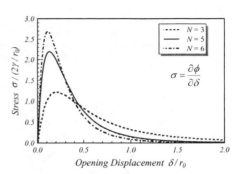

$$\sigma = \frac{\partial \phi}{\partial \delta}$$

Fig. 3　Relationship between crack opening displacement and bonding stress.

$$\phi_b(\delta_n) = \begin{cases} \dfrac{1}{2} \cdot K \cdot \delta_n^2 & (\delta_n \le 0) \\ 0 & (\delta_n \ge 0) \end{cases} \tag{3}$$

Where, δ_n and δ_t are the opening and shear deformation at the interface, respectively. The constants γ, r_0, and N are the surface energy per unit area, the scale parameter and the shape parameter of the potential function. In order to prevent overlapping in the opening direction due to a numerical error in the computation, the second term in Eq. (1) was introduced and K was set to have a large value as a constant. Also, to model an interaction between the opening and the shear deformations, a constant value A was employed in Eq. (2). From the above equations, the maximum bonding stress, σ_{cr}, under only the opening deformation δ_n and the maximum shear stress, τ_{cr}, under only the shear deformation δ_t are calculated as follows.

$$\sigma_{cr} = \frac{4\gamma N}{r_0} \cdot \left\{ \left(\frac{N+1}{2N+1} \right)^{\frac{N+1}{N}} - \left(\frac{N+1}{2N+1} \right)^{\frac{2N+1}{N}} \right\} \tag{4}$$

$$\tau_{cr} = \frac{4\gamma N \sqrt{A}}{r_0} \cdot \left\{ \left(\frac{N+1}{2N+1} \right)^{\frac{N+1}{N}} - \left(\frac{N+1}{2N+1} \right)^{\frac{2N+1}{N}} \right\} \tag{5}$$

By arranging such interface elements along the crack propagation path as shown in Fig. 2, the growth of the crack under the applied load can be analyzed in a natural manner. In this case, the decision on the crack growth based on the comparison between the driving force and the resistance as in the conventional methods is avoided. The parameters involving above two equations were determined for demonstrating α-Fe fracture behavior. Namely, the surface energy γ and the shape parameter N were assumed to 2.0 N/mm and 4, respectively[7]. The scale parameter r_0 is the property related to maximum bonding and sliding stresses (σ_{cr} and τ_{cr}), and was determined according to the misorientation angle between neighbor grains as described in the following section.

Anisotropy in Grain

The properties of each grain in metallic materials have anisotropy due to the direction of grain growth. In order to demonstrate the microstructural fracture behavior precisely, the anisotropy has to be taken into account and two methods were studied in this research. One was the method assuming the anisotropy of elastic moduli where the target plane was set to (110) plane since the α-Fe has hexagonal crystal lattice and the anisotropy of moduli is significantly appeared in (110) plane. Namely, the moduli of each direction in (110) plane (E_{11}, E_{22} and G_{12}) were assumed to 125.0, 210.5 and 66.20 GPa, respectively[8].

In another method, the theory of crystal plasticity based on the viscoplastic single crystal deformation was employed instead of the anisotropic moduli[9]. Namely, the slipping rate $\dot{\gamma}^{(\alpha)}$ of the αth slip system in a rate-dependent crystalline solid was defined by the corresponding resolved shear stress $\tau^{(\alpha)}$, and a simple form for the self hardening moduli $h(\gamma)$ was employed as for the hardening of rate-dependent crystalline materials according to the following equations.

$$\dot{\gamma}^{(\alpha)} = \dot{a}^{(\alpha)} \frac{\tau^{(\alpha)}}{g^{(\alpha)}} \left| \frac{\tau^{(\alpha)}}{g^{(\alpha)}} \right|^{n-1} \tag{6}$$

$$h(\gamma) = h_0 \operatorname{sech}^2 \left| \frac{h_0 \gamma}{\tau_s - \tau_0} \right| \tag{7}$$

Where $\dot{a}^{(\alpha)}$, $g^{(\alpha)}$ and n were the reference strain rate on slip system α, a variable describing current strength of that system and the rate sensitivity exponent, respectively. Meanwhile h_0, τ_0, τ_s and γ were the initial hardening modulus, the yield stress which equals the initial value of current strength $g^{(\alpha)}$, the stage I stress and the Taylor cumulative shear strain on all slip systems, respectively.

Anisotropy at Grain Boundary

Generally, the intergranular fracture is caused by crack opening and slipping at grain boundaries. The interface element can demonstrate such deformations by employing the interaction parameter A in Eq. (2). In the case with small value of parameter A, the interfacial slipping easily occurs in comparison with the crack opening. The parameter A was varied from 0.01 to 1.0 in order to examine the influence of difference between opening and shear strength (σ_{cr} and τ_{cr}) on the microstructural fracture behavior.

The grain boundary can be regarded as disorder of atoms between neighbor grains. So, it can be assumed that the fracture strength at grain boundary would be related to the boundary energy, which could be determined by the atomic disorder at the boundary. There have been many researches about the symmetric tilt boundary, and various boundary energies in different plane were studied by using the molecular dynamic method. Tanaka et al. examined <110> boundary energy of molybdenum and revealed that the square of fracture strength σ_{gb} would be linearly proportional to the grain boundary energy γ_{bd}[10]. Since the boundary energy of α-Fe was reported as shown in Fig. 4[11], the authors assumed that the opening fracture stress of α-Fe would also follow the relation to misorientation as shown in Fig. 5.

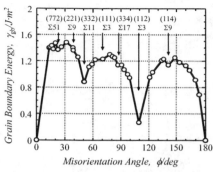

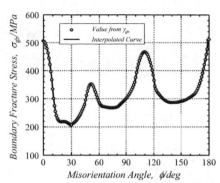

Fig. 4   Relationship between grain boundary energy and misorientation angle.

Fig. 5   Relationship between boundary fracture stress and misorientation angle.

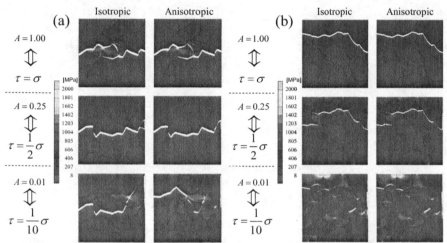

Fig. 6   Influence of anisotropy in grain and at grain boundary on microstructural fracture behavior (a) Model 30-1, (b) Model 30-2.

RESULTS AND DISCUSSIONS

Elastic Analysis

As a basic research for examining the applicability of this method for the microstructural fracture process, the influences of anisotropy in grain and interaction parameter A on the fracture behavior were studied assuming only the elastic deformation in grain while the grain boundary was set to anisotropic due to the neighbor grain orientations.   Figure 6 shows typical results of isotropic and anisotropic grains in Model 30-1 under applying 3.0 % strain and in Model 30-2 under applying 2.0 % strain, where the parameter A was varied and the tensile stress distributions were plotted.   In only the cases of Model 30-1 with 0.01 for A, the effect of anisotropy in grain on the crack propagation

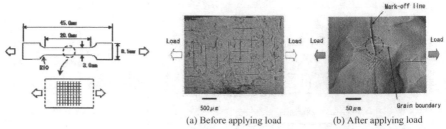

(a) Before applying load   (b) After applying load

Fig. 7   Experimental procedure for measuring microstructural fracture behavior.

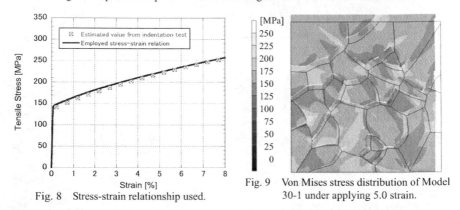

Fig. 8   Stress-strain relationship used.

Fig. 9   Von Mises stress distribution of Model 30-1 under applying 5.0 strain.

behavior could be observed while the fracture behaviors in other cases would not be affected by the anisotropy in grain.   On the other hand, the influence of parameter A was significantly observed when A was 0.01 regardless of grain configuration since the slipping at grain boundary was generated. From these results, it can be concluded that the interaction parameter A would be a dominant effect on the microstructural fracture behavior.

Elastic-Plastic Analysis

In order to examine the applicability of this method for the elastic-plastic problems, the slip deformation at grain boundary were calculated and the computed results were compared with the experimental observations conducted by Dr. Higuchi[6], where the mechanical property of grain was assumed to isotopic according to the elastic analyses.   Figure 7 shows the experimental procedures. In the experiment, the grid mark-off lines were marked on the surface of tensile specimen for ferritic steel and the tensile load was applied.   By applying the tensile load, the slip deformations at grain boundary could be obtained through the microscopic observation of the grain boundary across the mark-off lines.   The slip lengths parallel and transverse to the loading direction were measured separately.   On the other hand, in the numerical analyses, five lines parallel and transverse to the direction of forced displacement were set for examining the interfacial slip at grain boundary and the computed slip lengths were compared with the experimental results.   According to the experimental indentation tests of the grains, the grain was assumed to the elastic-exponential work hardening plastic material where initial yield stress, strength index and strain hardening index were set to 125 MPa, 45,

Table 1   Effect of interaction parameter o slip length at grain boundary.

Direction of slip to the loading direction	Experiments	Computations			
		$\sqrt{A}$ =0.10	$\sqrt{A}$ =0.25	$\sqrt{A}$ =0.50	$\sqrt{A}$ =1.00
Parallel	2.94 μm	4.97 μm	2.53 μm	1.50 μm	0.67 μm
Transverse	1.83 μm	1.88 μm	1.17 μm	0.44 μm	0.01 μm

Table 2   Slip length at grain boundary before excluding minimum values measured.

Direction of slip to the loading direction	Model 50-1	Model 50-2	Model 30-1	Model 30-2
Parallel	2.46 μm	2.80 μm	2.53 μm	2.78 μm
Transverse	1.20 μm	1.06 μm	1.17 μm	1.24 μm

Table 3   Slip length at grain boundary after excluding minimum values measured.

Direction of slip to the loading direction	Model 50-1	Model 50-2	Model 30-1	Model 30-2
Parallel	3.07 μm	3.19 μm	3.25 μm	3.30 μm
Transverse	1.94 μm	1.86 μm	1.76 μm	1.66 μm

0.3, respectively (Fig. 8). Meanwhile the interfacial property at grain boundary was assumed to anisotropic due to the neighbor grain orientations as same as the elastic analysis.

As a typical computational result, the von Mises stress distribution of Model 30-1 under applying 5.0 % strain is shown in Fig. 9 and the crack propagation behavior could be demonstrated. Table 1 shows the experimental and computational results of Model 30-1 where the square root of interaction parameter $\sqrt{A}$ was varied from 0.01 to 1.00 in order to study the influence of interfacial property on the microstructural fracture behavior. From this result, it was found that an appropriate value of $\sqrt{A}$ for demonstrating the interfacial behavior would be 0.25. The same analyses were conducted using other models shown in Fig. 1

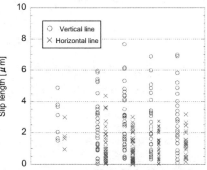

Fig. 10   Slip length at grain boundary measured and computed.

assuming the value of $\sqrt{A}$ as 0.25, all the values of the slip lengths parallel and transverse to the direction of forced displacement computed were plotted into Fig. 10 and the average values were summarized into Table 2. These results indicated that the simulation results had a good agreement with the experiments regardless of model. However, since the computational results contains small values which might not be identified in the experiment, the values less than the minimum value measured were excluded and the modified average slip lengths were summarized into Table 3. From

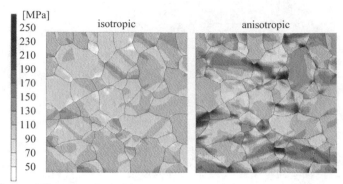

Fig. 11   Effect of mechanical property in grain on von Mises stress distributions.

this table, it can be concluded that the slipping behavior at grain boundary might not be affected by the grain configuration and the anisotropy of interfacial property would be a dominant factor.

As for another potentiality of FEM with the interface element, the influence of crystal plasticity on the microstructural deformation was examined by using Model 50-1 as shown in Fig. 1 instead of the elastic anisotropy of grain.   The mechanical property at grain boundary was assumed to independent of the neighbor grain orientations in order to point up the effect of crystal plasticity.   The mechanical property of grain was defined by Eqs. (6) and (7), where $\dot{a}^{(\alpha)}$, n, h_0, τ_0 and τ_s were set to 0.001 /s, 10, 500 MPa, 50 MPa and 100 MPa according to the previous study[9].   As the object of a comparison, the computation assuming the elastic-exponential work hardening plastic material as the isotropic grain was also conducted, where initial yield stress, strength index and strain hardening index were set to 95 MPa, 30, 0.3, respectively in order to fit the apparent strain-stress relationship of virtual model with the crystal plasticity to that of the isotropic grain.   While Young's modulus and Poisson's ratio were assumed to 200 GPa and 0.3, respectively for both cases.   Von Mises stress distributions under applying 5.0 % strain were summarized into Fig. 11.   Although any significant openings at grain boundary could not be obtained in both cases, the stress concentrations near the triple point at grains could be observed in the analysis with assuming the crystal plasticity.   Since the interfacial slipping was identified through the detailed analysis near the triple point, the slipping would enhance the stress concentration near the triple point.   Namely, it was revealed that the stress concentrations caused by both the mismatch between neighbor grains and the slipping at the grain boundary could be demonstrated by using FEM with the interface element.   Moreover, it can be concluded that this FEM with the interface element would be a useful tool for examining microstructural fracture behavior.

CONCLUSIONS

A finite element method with the interface element was developed for examining the microstructural fracture behavior in the multifunctional materials and its applicability was studied through the numerical analyses using virtual polycrystalline models.   From the results assuming the anisotropic property at grain boundary caused by the mismatch between the neighbor grain orientations, it was found that the interaction between opening and slipping deformations at grain boundary would be a dominant factor of the microstructural fracture process.   Also, through the analysis considering the crystal plasticity, it was revealed that the interfacial slipping would enhance the stress concentration near the triple point which was produced by the misorientation angle between the neighbor grains.   Moreover, the finite element method with the interface element would have a good potential for examining the microstructural fracture behavior.

ACKNOWLEDGEMENTS
The authors would like to express his sincere appreciation to Dr. R. Higuchi (Sumitomo Metal Industries, Ltd.) for providing the details of experimental result and for fruitful discussion.

REFERENCES
[1] H. Murakawa, H. Serizawa and Z.Q. Wu, "Computational Analysis of Crack Growth in Composite Materials Using Lennard-Jones Type Potential Function," *Ceramic Engineering and Science Proceedings*, **20** [3], 309-316 (1999).
[2] H. Serizawa, H. Murakawa and C.A. Lewinsohn, "Modeling of Fracture Strength of SiC/SiC Composite Joints by Using Interface Elements," *Ceramic Transactions*, **144**, 335-342 (2002).
[3] H. Serizawa, C. A. Lewinsohn, M. Singh and H. Murakawa, "Evaluation of Fracture Behavior of Ceramic Composite Joints by Using a New Interface Potential", *Materials Science Forum*, **502**, 69-74 (2005).
[4] H. Serizawa, D. Fujita, C. A. Lewinsohn, M. Singh and H. Murakawa, "Finite Element Analysis of Mechanical Test Methods for Evaluating Shear Strength of Ceramic Composite Joints Using Interface Element", *Ceramic Engineering and Science Proceedings*, **27** [2], 115-124 (2006).
[5] H. Serizawa, T. Hajima, S. Tomiyama and H. Murakawa, "Development of Numerical Method for Evaluating Microstructural Fracture in Smart Materials", *Ceramic Engineering and Science Proceedings*, **32** [8], 133-140 (2011).
[6] R. Higuchi, "Proposal of FEM-MD Coupling Simulation Method for Microscopic Stress Analysis Considering Deformation near Grain Boundary in Polycrystalline Steels", *Doctoral Thesis at Osaka University*, 1998.
[7] H. Murakawa, H. Serizawa, T. Tsujimoto and S. Tomiyama, "Mechanism and Effect of Stress-induced Transformation on Improvement of Fracture Toughness", *Transactions of JWRI*, **38**, 2, 71-78 (2009).
[8] N. Igata, "Strength of Materials", *BAIFUKAN Co., Ltd.*, Tokyo, 1983.
[9] Y. Huang, "A User-Material Subroutine Incorporating Single Crystal Plasticity in the ABAQUS Finite Element Program", Mech Report 178, *Division of Engineering and Applied Sciences, Harvard University*, Cambridge, Massachusetts, 1991.
[10] T. Tanaka, S. Tsurekawa, H. Nakashima and H. Yoshinaga, "Misorientation Dependence of Fracture-Stress and Grain-Boundary Energy in Molybdenum with [110] Symmetrical Tilt-Boundaries", *Journal of the Japan Institute of Metals*, **58**-4, 382-389 (1994).
[11] H. Nakashima and M. Takeuchi, "Grain Boundary Energy and Structure of α-Fe(110) Symmetric Tilt Boundary", *TETSU-TO- HAGANE*, **86**-5, 357-362 (2000).

Author Index